Turkish Foreign Policy in an Age of Uncertainty

F. Stephen Larrabee • Ian O. Lesser

Prepared for the

Center for Middle East Public Policy

NATIONAL SECURITY RESEARCH DIVISION

RAND

The research described in this report was prepared for RAND's Center for Middle East Public Policy.

Library of Congress Cataloging-in-Publication Data

Larrabee, F. Stephen.
    Turkish foreign policy in an age of uncertainty / F. Stephen Larrabee, Ian O. Lesser.
        p. cm.
    "MR-1612."
    Includes bibliographical references.
    ISBN 0-8330-3281-X
    1. Turkey—Foreign relations—1980– I. Lesser, Ian O., 1957– II.Title.

DR477 .L37 2002
357.561'09'0511—dc21

2002015207

RAND is a nonprofit institution that helps improve policy and decisionmaking through research and analysis. RAND® is a registered trademark. RAND's publications do not necessarily reflect the opinions or policies of its research sponsors.

Cover design by Stephen Bloodsworth

Published 2003 by RAND
1700 Main Street, P.O. Box 2138, Santa Monica, CA 90407-2138
1200 South Hayes Street, Arlington, VA 22202-5050
201 North Craig Street, Suite 202, Pittsburgh, PA 15213-1516
RAND URL: http://www.rand.org/
To order RAND documents or to obtain additional information, contact Distribution Services: Telephone: (310) 451-7002;
Fax: (310) 451-6915; Email: order@rand.org

Not since the end of the Cold War, and perhaps not since the early days of the Turkish Republic, has Turkey's external role been characterized by so many open questions. What place for Turkey in Europe? What risks and opportunities for Turkey in a conflict-ridden Middle East? How will Ankara deal with a changing Russia, an unstable Caucasus, and Central Asia? To what extent can a traditionally competitive relationship with Greece be moderated, against a background of successive Balkan crises? Turkey may be a pivotal state in Western perception, but uncertainties in transatlantic relations may make the very concept of the "West" unclear as seen from Ankara. Above all, Turkey faces daunting political, economic, and social pressures, with implications for the vigor and direction of the country's foreign and security policies. The range of possibilities is now quite wide, from a more globalized Turkey, more closely integrated in Europe and the West, with a multilateral approach toward key regions, to a more inward-looking and nationalistic Turkey, pursuing a more constrained or unilateral set of regional policies.

This book seeks to describe the challenges and opportunities facing Turkey in the international environment at a time of extraordinary flux and offers some conclusions about the country's future. The analysis should be of interest to policymakers and observers in Turkey, in the West, and elsewhere and contribute to informed debate about the country's role. The volume builds on a substantial body of RAND work on Turkey and related topics undertaken over the last decade. This study extends and updates this tradition of analysis. In addition to exploring the internal and regional dimensions of Turkish foreign policy, we have given special emphasis throughout to the strategic and security issues facing Turkey. These include a number of new issues posed by the terrorist attacks on the United States of September 2001 and the looming confrontation with Iraq.

The research for this study was conducted within RAND's Center for Middle East Public Policy. The center takes an integrative approach to policy studies, applying RAND expertise in strategic and defense analyses, science and technology, human capital, and other areas to the complex issues facing the Middle East and nearby regions.

CONTENTS

SUMMARY

As Turkey enters the 21st century, it faces a troubled environment, domestically and internationally. Uncertainties regarding the country's future and its external policies have increased significantly as a result of Turkey's own economic crises and political turmoil, troubling developments in nearby regions, and challenges further afield. As a consequence, the task of understanding and assessing Turkey's international role has become more complex and far more difficult.

During the Cold War, Turkey acted as a bulwark against the expansion of Soviet influence into the Eastern Mediterranean and Middle East. With the end of the Cold War, Ankara's policy horizons have expanded and Turkey has become a more assertive and independent actor on the international stage. Where once Turkey primarily looked West, today Turkey is increasingly being pulled East and South as well. As a result, Turkey has been forced to redefine its foreign and security policy interests and to rethink its international relationships.

If Turkey were a small state located in Antarctica or the South Sea Islands, these changes might matter little. But Turkey stands at the nexus of three areas of increasing strategic importance to the United States and Europe: the Balkans, the Caspian region, and the Middle East. Thus, how Turkey evolves is important, both to the United States and to Europe.

Turkey's sheer size, moreover, gives it important geostrategic weight. Turkey's population is currently nearly 68 million—the second largest in Europe behind Germany—and may be close to 100 million by the middle of the 21st century. This would make Turkey the most populous country in Europe. Integrating a country and economy of this size will place significant burdens on a European Union (EU) already reeling from the demands posed by admitting much smaller countries from Central and Eastern Europe. The challenges for Turkey and Europe will be daunting. How each side responds to these challenges will have an important effect not only on Turkey's evolution but on Europe's political and strategic evolution as well.

TURKEY AS A REGIONAL ACTOR

In the past decade, moreover, Turkey has emerged as an increasingly important regional actor, wielding substantial military as well as diplomatic weight. Nowhere has this been more evident than in the Middle East. This growing involvement in Middle Eastern affairs represents an important shift in Turkish policy. Under Atatürk—and for several decades after his death—Turkey eschewed involvement in the Middle East, but in recent years, Turkey has been heavily engaged in the region. This more active policy contrasts markedly with the more passive approach that characterized Turkish policy before the Gulf War.

At the same time, Turkey's greater involvement in the Middle East has complicated relations with Europe. Many Europeans are wary of Turkish membership in the EU not only because of the political, economic, and cultural problems it would present but also because they fear it will extend Europe's borders into the Middle East and drag Europe more deeply into the vortex of Middle Eastern politics. Thus, Turkey's Middle Eastern involvement has raised new dilemmas about its European or Western identity. The deeper its involvement in the Middle East, the more problems this poses for Turkey's Western orientation and identity.

The end of the Cold War has also opened up new horizons for Turkish policy in the Caucasus and Central Asia—areas that were previously closed to Turkish policy. Although Turkey has been cautious about exploiting these possibilities, the emergence of the Caucasus and Central Asia has given a new dimension to Turkish policy. Turkey now has interests in the region that it did not have during the Cold War. This inevitably affects its security perceptions and relations with its Western allies.

At the same time, Turkey's interest and involvement in the Caucasus and Central Asia have complicated relations with Russia and given the historical rivalry between the two countries new impetus. Increasingly, Russia has come to see Turkey as a major rival for influence in the region and has sought to stem Turkey's efforts to establish a geostrategic foothold there. But Russia also remains an increasingly important trade partner. This gives Turkey a strong incentive to keep relations with Russia on an even keel. Indeed, the growing economic interaction between Russia and Turkey is one of the most important developments in Turkish policy toward Eurasia and could have a significant effect on Turkey's relations with Moscow over the long run.

Turkey's relations with Europe are also undergoing important and stressful change. At its December 1999 Helsinki summit, the EU decided to accept Turkey as a candidate for membership, after years of keeping it at arm's length. Since then, Turkey has undertaken a number of important reforms designed to meet the EU's Copenhagen criteria, including abolition of the death penalty and an easing of restrictions on the use of the Kurdish language. However, many Europeans are still not convinced that Turkey should be admitted, both for cultural as well as economic reasons; their reservations are compounded by the sheer scale of Turkey as a society.

Perhaps the area where Turkey's relations have witnessed the most dramatic change, however, is with Greece. After years of hostility, Greek-Turkish relations have slowly begun to improve, bolstered in part by "earthquake diplomacy." The key question, however, is whether the recent rapprochement simply represents a tactical maneuver or fundamental strategic change in the nature of relations. So far, the thaw has been limited largely to nonstrategic areas such as trade, the environment, tourism, and a variety of nontraditional security matters. However, if it is to be durable, it will need to address the core issues of the Aegean and Cyprus.

In the Balkans, too, Turkish policy is in flux. After the collapse of the Ottoman empire, Turkey effectively withdrew from the Balkans. But the end of Cold War has witnessed renewed Turkish interest in the region. Turkey's relations with Albania, Macedonia, and Bulgaria have visibly improved. Turkey has actively participated in peacekeeping and stabilization operations in Bosnia and Kosovo and would likely contribute to any Western peacekeeping operation in Macedonia. But Turkey's sympathy for the Muslims in Bosnia and elsewhere worries many Europeans—especially Greeks—who fear that at some point Turkey might be tempted "to play the Muslim card." So far, Ankara's approach to the region has been moderate and multilateral. But a more nationalist government in Ankara might not be as restrained.

Finally, Turkey's relations with the United States have witnessed important changes. Turkey's increasing involvement in the Balkans, the Caucasus, and the Middle East have increased Turkey's strategic importance in Washington's eyes. The United States has come to see Turkey as a key strategic ally and a more capable actor in these regions. In addition, the war on terrorism, and the U.S. desire to bolster moderate voices in the Muslim world, have reinforced Turkey's strategic importance to the United States.

But U.S.-Turkish perspectives differ on many issues, especially in the Gulf. Turkey has strong reservations about U.S. policy toward Iraq, which it fears will lead to the creation of a separate Kurdish state in Northern Iraq. A U.S. invasion of Iraq could put new strains on U.S.-Turkish relations. Ankara also does not share Washington's view about the need to isolate Iran, which is an important trading partner and a source of natural gas for Turkey. These differences hinder the development of a true "strategic partnership" between Turkey and the United States.

THE TRANSNATIONAL DIMENSION

Turkey's interests and policy are also shaped by a number of cross-cutting transnational—and transregional—issues, especially the spread of weapons of mass destruction (WMD) and the proliferation of ballistic missile technology. Turkey's increased exposure to WMD is bound to influence its security perceptions in the future. At the same time, this exposure gives Turkey a much stronger interest in missile defense than many of its European allies, which do not (yet) face the same degree of vulnerability to these threats.

Terrorism is another transnational issue having a significant effect on Turkish security perceptions. The number of assassinations of prominent Turkish officials and journalists in the last decade, and the persistence of left-wing and right-wing terrorism inside Turkey, have heightened Turkish sensitivity to the dangers of international terrorism and given Ankara a strong interest in combating its spread. This could become an issue in Turkey's relations with the United States, especially if some of Turkey's neighbors, or groups within Turkey, were to begin to conduct terrorist attacks against Turkish bases from which U.S. forces were operating.

Energy has also emerged as an important factor influencing Turkish policy. Turkey's growing energy needs have given Turkey a strong interest in developing ties to energy-producing states in the Middle East and the Caspian region. Turkey's strong support for the development of the Baku-Ceyhan pipeline has become a critical element in its strategy in the Caucasus and Caspian region. Turkey has also sought to expand economic cooperation with Iran in the energy field despite strong U.S. objections. And it has been eager to revive energy cooperation with Iraq and see the current sanctions regime against Baghdad lifted. With the development of new oil and gas routes to bring Caspian and Middle Eastern supplies to world markets, Turkey is also emerging as a key energy entrepôt and transshipment state, especially for Europe.

Finally, Turkish policy has been affected by the increased emphasis on human rights in Western policy. The Kurdish issue, in particular, has been a source of tension in Turkey's relations with Europe, especially Germany. But human rights concerns have also had an important effect on relations with the United States. Concerns in this area have had a substantial influence on the character of the bilateral relationship in recent years, including security cooperation and arms transfers.

INTERNAL CHANGE

These changes are occurring at a time when the Turkish domestic scene is also experiencing important changes, many of which are having an effect on Turkey's foreign and security policy. In the last decade, many of the key tenets of Atatürkism—Westernization, statism, secularism, and nonintervention—have come under increasing assault. The democratization of Turkish society has created space for a variety of new groups and forces that have challenged the power of the Kemalist state. These challenges, including challenges from Turkey's secular reformists to the traditionally strong state apparatus, will be a key force shaping Turkish society and policy over the next decades.

In addition, the Islamist movement in Turkey has undergone an important evolution in the last several years. The dominant tendency among Turkey's religious politicians today—many operating under legal bans on political activity—is toward what is described as "Muslim democracy," loosely patterned on the model of Christian democratic parties in Europe. The foreign policy orientation of these Muslim democrats appears increasingly mainstream but reflects a degree of wariness regarding globalization and integration and can be nationalistic in tone.

Indeed, the reassertion of Turkish nationalism is arguably a far more important influence on foreign policy than religious politics in Turkey. Although nationalism has been a key component of Turkish foreign policy going back to Atatürk, the Gulf War gave it new impetus. Many Turks felt that Turkey paid too high an economic and political price for its support of the United States in the Gulf War. Moreover, these sacrifices did not bring the expected rewards, namely, membership in the European Union. The tepid European response provoked considerable resentment among the Turkish public and reinforced a sense that Turkey had to look after its own interests more vigorously. Events in Bosnia and Chechnya, where Turkish affinities are engaged, reinforced this nationalist inclination. In a very different fashion, Turkey's current economic crisis—and resentment over the role of international financial institutions—has also been a spur to nationalist sentiment.

Another important trend has been the growing influence of the independent media, especially television. The media played an important role in the crisis over Imia/Kardak, which brought Turkey and Greece to the brink of war in January 1996, as well as in the first Chechnya crisis. Moreover, Islamist and other political groups now have their own television stations, giving them unprecedented access to a much broader cross section of the Turkish public. Public opinion and the media are now far more important factors in Turkish external policy than ever before.

At the same time, the growth of a dynamic private sector has served to weaken the role of the "strong state" and strengthen the power of civil society. The business community, in particular, has emerged as an important political force in Turkey. The Turkish Industrialists' and Businessmen's Association and other institutions have emerged as outspoken advocates for reform and have proposed new policy initiatives on a broad range of social, economic, and political issues, including the Kurdish issue. Turkish entrepreneurs have also played a leading role in the expansion of economic ties to Russia, Central Asia, and the Middle East and have been at the forefront of the recent rapprochement with Greece. Together with the changing role of the military in Turkish society, the landscape for debate and policymaking on a range of issues, including foreign and security policy, is changing rapidly, with new actors operating alongside traditional elites.

Finally, the role of the Turkish military is evolving. The military has traditionally regarded itself as the custodian of the Atatürk legacy and has directly intervened three times when it felt democracy in Turkey was threatened. Each time, however, it has returned to the barracks after a short period of direct rule. Today, the military is much less inclined to intervene directly in Turkish politics. But it remains an important political force behind the scenes, as its ouster of the Erbakan government in a "silent coup" on June 1997 underscores. This political role is regarded by many EU members as inconsistent with a modern democracy and could pose an obstacle to Turkey's membership in the EU over the long term.

In sum, Turkey today is at an important crossroads. Externally, it faces new challenges, especially in Europe and the Middle East; internally, it has reached an impasse that requires important changes in the way Turkey is governed—and by whom. Incrementalism and "muddling through"—approaches that have characterized Turkish policy in the past—are unlikely to be sufficient in the future. The threshold for unrest in Turkey remains high, but a continuing economic crisis, with social and political cleavages left unresolved, could push Turkey toward greater instability, making extreme or more chaotic outcomes possible. Although far from inevitable, these outcomes would have serious implications for Turkey's external relations. In particular, a Turkey in turmoil would result in even more resistance to the idea of Turkish membership in the EU. Turkey would also become a much less dependable ally for the United States. Hence, how Turkey resolves these challenges matters—both for Turks and for Western policy.

ACKNOWLEDGMENTS

The authors have benefited enormously from recent discussions with public and private sector observers in Turkey, Europe, and the United States. We are most grateful to all who contributed views and comments. We also wish to thank RAND colleagues Daniel Byman, Rachel Swanger, Jeff Isaacson, and Nurith Berstein, as well as Barbara Kliszewski, Shirley Birch, and Chrystine Keener, for their assistance in the research for and preparation of the manuscript. Bülent Aliriza, Keith Crane, Theodore Couloumbis, Philip Gordon, Fiona Hill, Olga Oliker, Sabri Sayari, and Thanos Veremis offered valuable comments and advice on earlier drafts of the book. Finally, we owe special thanks to Jerrold Green and RAND's Center for Middle East Public Policy, and Stuart Johnson and James Dobbins of RAND's National Security Research Division, for their generous support of this project. Of course, any errors or omissions are the sole responsibility of the authors.

ABM	Antiballistic Missile
ACRS	Arms Control and Regional Security
AKP	Justice and Development Party
ANAP	Motherland Party
AP	Accession Partnership
BLACKSEAFOR	Black Sea Naval Task Force
BSEC	Black Sea Economic Cooperation
CENTO	Central Treaty Organization
CFE	Conventional Forces in Europe
DECA	Defense and Economic Cooperation Agreement
DSP	Democratic Left Party
DYP	True Path Party
EC	European Community
ESDI	European Security and Defense Initiative
ESDP	European Security and Defense Policy
EU	European Union
GDP	Gross Domestic Product

GNP	Gross National Product
ICJ	International Court of Justice
IFOR	Implementation Force (Bosnia)
IISS	International Institute for Strategic Studies
IMF	International Monetary Fund
ISAF	International Security Assistance Force (Afghanistan)
KFOR	Kosovo Peacekeeping Force
MHP	Nationalist Action Party
MUSIAD	Independent Turkish Businessmen's Association
NATO	North Atlantic Treaty Organization
NSC	National Security Council
OECD	Organization for Economic Development and Co-operation
OSCE	Organization for Security and Cooperation in Europe
PASOK	Pan-Hellenic Socialist Movement
PERMIS	Permanent International Secretariat
PKK	Kurdistan Workers Party
SEEBRIG	Southeast European Brigade
SFOR	Stabilization Force
TCGP	Trans-Caspian Gas Pipeline
TESEV	Turkish Economic and Social Studies Foundation
TGS	Turkish General Staff
TIKA	Turkish International Cooperation Agency

TRNC	Turkish Republic of Northern Cyprus
TUSIAD	Turkish Industrialists' and Businessmen's Association
WEU	Western European Union
WMD	Weapons of Mass Destruction
YTP	New Turkey Party

Chapter One

INTRODUCTION: TURKISH FOREIGN POLICY IN TRANSITION

Turkey faces a troubled environment, domestically and internationally. Uncertainties regarding the country's future and its external policies have increased significantly as a result of Turkey's own economic crises and political turmoil, troubling developments in nearby regions, and challenges further afield. The opening of the 21st century has seen a multiplication of variables influencing Turkey's foreign and security policy. As a consequence, the task of understanding and assessing Turkey's international role has become more complex and far more difficult.

During the Cold War, Turkey was a key part of the Western defense system. Ankara acted as a bulwark against the expansion of Soviet influence into the Eastern Mediterranean and Middle East. It tied down some 24 Soviet divisions that might otherwise have been deployed on the Central front. It also supplied important bases and facilities for the forward deployment of nuclear weapons and the monitoring of Soviet compliance with arms control agreements. With the end of the Cold War, many in Turkey and the West assumed a much reduced role for Turkey as a regional actor and as an ally of the West. These assumptions, however, proved unfounded. Rather than declining, Turkey's strategic importance has increased.

At the same time, Ankara's policy horizons have expanded and Turkey has become a more assertive and independent actor on the international stage. Where once Turkey primarily looked West, today Turkey is increasingly being pulled East and South as well. As a re-

sult, Turkey has been forced to redefine its foreign and security policy interests and to rethink its international relationships.

At the same time, Turkey has faced new domestic challenges from
Kurdish separatists and Islamists. Economic changes, especially the
growth of a dynamic private sector, have eroded the role of the state
and created new political and economic forces that have challenged
the power of the old Kemalist bureaucratic elite. These forces have
increasingly influenced both the style and substance of Turkish foreign and security policy. Indeed, the debate between state-centered
conservatives and reformers—a debate that cuts across private and
public circles in Turkish society—has been greatly sharpened by the
economic crisis of 2000–2002. Looking ahead, the crisis and its political consequences could have important implications for the composition and orientation of Turkey's foreign policy elite.

TURKEY AS A "PIVOT" STATE

If Turkey were a small state located in Antarctica or the South Sea Is-
lands, these changes might matter little. But Turkey stands at the
nexus of three areas of increasing strategic importance to the United
States and Europe: the Balkans, the Caspian region, and the Middle
East. Thus, how Turkey evolves is important both to the United
States and to Europe.

Indeed, in many ways, Turkey is a "pivot" state par excellence.[1]
Population, location, and economic and military potential are key
requirements for pivot states. But the defining quality of a pivot state
is, above all, the capacity to affect regional and international stability.
By this measure, Turkey clearly qualifies, along with such states as
Mexico, Brazil, Algeria, Egypt, India, and Indonesia. This disparate
list of states is tied together by their capacity to promote regional
stability—or disorder. A prosperous, stable Turkey would be a factor
for stability in a number of different areas: the Balkans, the Caucasus, the Middle East, and Europe. But an impoverished, unstable

---

[1]On the definition and role of a pivot state, see Robert S. Chace, Emily Hill, and
Paul Kennedy, "Pivotal States and U.S. Strategy," *Foreign Affairs*, Vol. 75, No. 1,
January/February 1996, pp. 33–51.

Turkey wracked by religious, ethnic, and political turmoil would be a source of instability and concern in all four regions.

What sets Turkey apart from other developing pivot states is its membership in the Western strategic club, principally through the North Atlantic Treaty Organization (NATO) but also through its evolving relations with the European Union (EU). Thus, developments in Turkey are directly linked to U.S. and Western interests. A reorientation of Turkey's foreign policy or serious threat to its democratic order would have important political and security consequences for both the United States and Europe.

Turkey's sheer size, moreover, gives it important geostrategic weight. Turkey's population is currently nearly 67.8 million—the second largest in Europe behind Germany—and may be close to 100 million by the middle of the 21st century. This would make Turkey the most populous country in Europe. Integrating a country and economy of this size will place significant burdens on an EU already reeling from the demands posed by admitting much smaller countries from Central and Eastern Europe. The challenges for Turkey and Europe will be daunting. How each side responds to these challenges will have an important effect not only on Turkey's evolution but on Europe's political and strategic evolution as well.

TURKEY AS A REGIONAL ACTOR

In the past decade, moreover, Turkey has emerged as an increasingly important regional actor, wielding substantial military as well as diplomatic weight. Nowhere has this been more evident than in the Middle East. This growing involvement in Middle Eastern affairs represents an important shift in Turkish policy. Under Atatürk—and for several decades after his death—Turkey eschewed involvement in the Middle East, but in recent years Turkey has been heavily engaged in the region. The Gulf War was an important turning point in this process. Against the counsel of his security advisors, President Özal opted squarely to allow the United States to fly sorties against Iraq from Turkish bases. Turkey also shut down the Kirkuk-Yumurtalik oil pipeline as part of the effort to impose economic sanctions against Iraq.

Özal's action was an important departure from Turkey's traditional policy of avoiding deep involvement in Middle Eastern affairs. At the same time, it opened a new period of greater activism in Turkish policy toward the Middle East, which has intensified visibly since the mid-1990s. This more active policy contrasts markedly with the more passive approach that characterized Turkish policy before the Gulf War.

The most dramatic example of this new approach to the Middle East has been Turkey's growing relationship with Israel. The Israeli connection has strengthened Turkey's diplomatic leverage in the region and was a factor in Ankara's decision to force a showdown with Syria in the Fall of 1998 over Syria's support for the Kurdistan Workers Party (PKK). The renewal of Arab-Israeli tensions, however, could put new strains on Turkey's relations with Israel.

Deeper involvement in the Middle East has not been cost free. The burgeoning security relationship with Israel has complicated Turkey's already mixed relations with its Arab neighbors. Turkey also faces new threats, including from weapons of mass destruction (WMD) deployed on or near its borders. Turkey is already within range of ballistic missiles that could be launched from Iran, Iraq, and Syria and this exposure is likely to grow in the future as more countries in the region acquire ballistic missile technology and the capability to deploy weapons of mass destruction. A nuclear-armed Iran or Iraq could dramatically change the security equation for Turkey and could have broader consequences for military balances elsewhere on Turkey's borders. The renewed confrontation between Israel and the Palestinians, counterterrorist operations in Afghanistan and possibly elsewhere, and the potential for conflict with Iraq place all these issues in sharper relief.

At the same time, Turkey's greater involvement in the Middle East has complicated relations with Europe. Many Europeans are wary of Turkish membership in the EU not only because of the political, economic, and cultural problems it would present, but because they fear it will extend Europe's borders into the Middle East and drag Europe more deeply into the vortex of Middle Eastern politics. Thus, Turkey's Middle Eastern involvement has raised new dilemmas about its European or Western identity. The deeper its involvement in the Middle East, the more problems this poses for Turkey's Western orientation and identity.

The end of the Cold War also opened up new horizons for Turkish policy in the Caucasus and Central Asia—areas that were previously closed to Turkish policy. Although Turkey has been cautious about exploiting these possibilities, the emergence of the Caucasus and Central Asia has given a new dimension to Turkish policy. Turkey now has interests in the region that it did not have during the Cold War. This inevitably affects its security perceptions and relations with its Western allies.

At the same time, Turkey's interest and involvement in the Caucasus and Central Asia have complicated relations with Russia and given the historical rivalry between the two countries new impetus. Increasingly, Russia has come to see Turkey as a major rival for influence in the region and has sought to stem Turkey's efforts to establish a geostrategic foothold there. But Russia also remains an increasingly important trade partner. This gives Turkey a strong incentive to keep relations with Russia on an even keel. Indeed, the growing economic interaction between Russia and Turkey is one of the most important developments in Turkish policy toward Eurasia and could have a significant effect on Turkey's relations with Moscow over the long run.

Turkey's relations with Europe are also undergoing important and stressful change. The EU's decision at the December 1999 Helsinki summit to accept Turkey as a candidate for membership, after years of keeping Turkey at arm's length, helped to ease strains in Turkey's relations with Europe. But membership would require major changes in Turkey's internal policies and practices. Turkey would have to cede a degree of sovereignty that many Turks may find difficult to accept. Membership could also require Turkey to open up its internal practices to outside scrutiny to an unprecedented degree, and the military will have to accept a less prominent role in Turkish politics. Several years after Helsinki, Europeans still have reservations about the pace of Turkish reform, and Turkish opinion about the EU has become more ambivalent and critical. Moreover, the prospects for Turkish progress on EU-related reforms are now closely tied to the outcome of Turkey's efforts to emerge from its economic and political crisis.

In short, Turkish membership is by no means assured. Many Europeans are still not convinced that Turkey should be admitted, both for cultural as well as economic reasons—reservations that are compounded by the sheer scale of Turkey as a society. At the same time, the EU's decision to create a distinct European Security and Defense Policy (ESDP) creates new security dilemmas for Turkey. Turkey is not a member of the EU and is not likely to be one in the near future. Thus, it continues to see NATO as the main vehicle for managing its security and defense problems. This sets it apart from many of its European allies who favor a stronger European and defense identity and complicates its relationship with Europe. As the EU develops a stronger security and defense component, the tensions between Turkey's strong attachment to NATO and its desire for EU membership could intensify.

Perhaps the area where Turkey's relations have witnessed the most dramatic change, however, is with Greece. After years of hostility, Greek-Turkish relations have slowly begun to improve, bolstered in part by "earthquake diplomacy." The key question, however, is whether the recent rapprochement is durable. Does the thaw represent a qualitative change in relations that will lead to a lasting détente or is it just the lull before a new storm? So far, the thaw has been limited largely to nonstrategic areas such as trade, the environment, tourism, and a variety of nontraditional security matters. However, if it is to be durable, it will need to address the core issues of the Aegean and Cyprus.

Cyprus, in particular, remains a major obstacle to a more far-reaching rapprochement. Indeed, if anything, Turkish views on Cyprus have hardened in recent years. Turkey has increasingly come to see Cyprus as a wider strategic issue, going beyond the protection of Turkish brethren on the island. This security dimension is likely to continue to color Turkish views on Cyprus and make any settlement of the issue difficult. With the EU's decision to admit Green Cyprus at its Copenhagen summit in December 2002 a near certainty, the Cyprus problem could become a flashpoint in relations between Ankara and Brussels.

In the Balkans, too, Turkish policy is in flux. After the collapse of the Ottoman empire, Turkey effectively withdrew from the Balkans. But the end of Cold War has witnessed renewed Turkish interest in the region. Turkey's relations with Albania, Macedonia, and Bulgaria have visibly improved. Turkey has actively participated in the Implementation Force (IFOR), the Stabilization Force (SFOR), and the Kosovo Peacekeeping Force (KFOR) and would likely contribute to any Western peacekeeping operation in Macedonia. But Turkey's sympathy for the Muslims in Bosnia and elsewhere worries many Europeans—especially Greeks—who fear that at some point Turkey might be tempted "to play the Muslim card." So far, Ankara's approach to the region has been moderate and multilateral. But a more nationalist government in Ankara might not be as restrained.

Finally, Turkey's relations with the United States have witnessed important changes. The United States has come to see Turkey as a key strategic ally and a more capable actor in these regions. Turkey's increasing involvement in the Balkans, the Caucasus, and the Middle East, in addition to the war on terrorism, and the U.S. desire to bolster moderate voices in the Muslim world, have reinforced Turkey's strategic importance in Washington's eyes.

But U.S.-Turkish perspectives differ on many issues, especially in the Gulf. Turkey has strong reservations about U.S. policy toward Iraq, which it fears will lead to the creation of a separate Kurdish state in Northern Iraq. A U.S. invasion of Iraq could put new strains on U.S.-Turkish relations. Ankara also does not share Washington's view about the need to isolate Iran, which is an important trading partner and source of natural gas for Turkey. These differences hinder the development of a true "strategic partnership" between Turkey and the United States.

Moreover, in recent years Turkey has become increasingly sensitive about issues of national sovereignty and has imposed tight restrictions on the use of its bases to monitor the no-fly zone over Northern Iraq. A more assertive Turkey—especially one more deeply involved in the Middle East—is likely to be even more sensitive about the use of its facilities in operations that directly affect its regional interests.

At the same time, Turkey's strategic environment will be strongly influenced by the evolution of America's regional and defense policies beyond the Gulf—including the new focus on counterterrorism worldwide. Washington has supported financial assistance to Ankara, as Turkey struggles to recover from its economic crisis, but tolerance for Turkey's renewed demands on the International Monetary Fund (IMF) may be reaching a limit. The character of U.S.-Russian relations will also have important implications for Ankara's own planning, and Turkey will have a strong stake in the evolution of U.S.-led missile defense efforts. More broadly, the evolution of U.S.-European relations will influence "where Turkey fits" in Western interests and strategy.

THE TRANSNATIONAL DIMENSION

Turkey's interests and policy are also shaped by a number of cross-cutting transnational—and transregional—issues, especially the spread of WMD and the proliferation of ballistic missile technology. Turkey's increased exposure to WMD is bound to influence its security perceptions in the future. At the same time, this exposure gives Turkey a much stronger interest in missile defense than many of its European allies, which do not (yet) face the same degree of vulnerability to these threats.

Terrorism is another transnational issue having a significant effect on Turkish security perceptions. The number of assassinations of prominent Turkish officials and journalists in the last decade, and the persistence of left-wing and right-wing terrorism inside Turkey, have heightened Turkish sensitivity to the dangers of international terrorism and given Ankara a strong interest in combating its spread. This could become an issue in Turkey's relations with the United States, especially if some of Turkey's neighbors, or groups within Turkey, were to begin to conduct terrorist attacks against Turkish bases from which U.S. forces were operating. Moreover, Turkey has defined its long struggle against the PKK insurgency in Southeastern Anatolia as a battle against terrorism, and the United States recognizes the PKK as a terrorist group. Turkey may also play a significant longer-term role in Western counterterrorism strategy in the Middle East and Central Asia.

Energy has also emerged as an important factor influencing Turkish policy. Turkey's growing energy needs have given Turkey a strong interest in developing ties to energy-producing states in the Middle East and the Caspian region. Turkey's strong support for the development of the Baku-Ceyhan pipeline has become a critical element in its strategy in the Caucasus and Caspian region. Turkey has also sought to expand economic cooperation with Iran in the energy field despite strong U.S. objections. And it has been eager to revive energy cooperation with Iraq and see the current sanctions against Baghdad lifted. With the development of new oil and gas routes to bring Caspian and Middle Eastern supplies to world markets, Turkey is also emerging as a key energy entrepôt and transshipment state, especially for Europe.

Finally, Turkish policy has been affected by the increased emphasis on human rights in Western policy. The Kurdish issue, in particular, has been a source of tension in Turkey's relations with Europe, especially Germany. But human rights concerns have also had an important effect on relations with the United States. Concerns in this area have had a substantial influence on the character of the bilateral relationship in recent years, including security cooperation and arms transfers. Human rights will undoubtedly remain on the agenda and will take on added significance as Ankara looks to European integration and also in light of Turkey's ambitious defense procurement plans. Indeed, one motivation behind Turkey's efforts to expand defense-industrial cooperation with Israel has been its desire to offset the human-rights-related constraints it has faced in trying to procure military equipment from the United States and key European countries such as Germany.

INTERNAL CHANGE

The Turkish domestic scene is also experiencing important changes, many of which are having an effect on Turkey's foreign and security policy. In the last decade, many of the key tenets of Atatürkism—Westernization, statism, secularism, and nonintervention—have come under increasing assault. The democratization of Turkish society has created space for a variety of new groups and forces that have challenged the power of the Kemalist state. These challenges, including challenges from Turkey's secular reformists to the traditionally strong state apparatus, will be a key force shaping Turkish society and policy over the next decades.

In addition, the Islamist movement in Turkey has undergone an important evolution in the last several years. The experience of an Islamist Refah-led government—forced to resign under pressure by the military and secular forces in June 1997—raised the specter of a changed foreign policy orientation, with greater attention to Turkey's relations with the Islamic world. In actual fact, there was little change in Turkish policy during this period, and even the Refah leadership was unwilling or unable to derail Turkey's expanding relationship with Israel. Refah's successor, the Fazilet (Virtue) Party, supported Turkish membership in NATO and actually championed the idea of EU membership on the assumption that European integration would mean more freedom of action for Turkey's Islamists. Today, with Fazilet banned, there is a tendency toward fragmentation in religious politics. The dominant tendency among Turkey's religious politicians—many operating under legal bans on political activity—is now toward what is described as "Muslim democracy," loosely patterned on the model of Christian democratic parties in Europe. The Justice and Development Party, the leading successor to the Virtue Party, enjoys widespread support among Turkey's electorate. The foreign policy orientation of these Muslim democrats appears increasingly mainstream but reflects a degree of wariness regarding globalization and integration and can be nationalistic in tone.

Indeed, the reassertion of Turkish nationalism is arguably a far more important influence on foreign policy than religious politics in Turkey. Although nationalism has been a key component of Turkish foreign policy going back to Atatürk, the Gulf War gave it new impetus. Many Turks felt that Turkey paid too high an economic and political price for its support of the United States in the Gulf War. Moreover, these sacrifices did not bring the expected rewards vis-à-vis membership in the European Union. The tepid European response provoked considerable resentment among the Turkish public and reinforced a sense that Turkey had to look after its own interests more vigorously. Events in Bosnia and Chechnya, where Turkish affinities are engaged, reinforced this nationalist inclination. In a very different fashion, Turkey's current economic crisis—and resentment over the role of international financial institutions—has also been a spur to nationalist sentiment.

The coalition that emerged after the 1999 elections, comprising Prime Minister Bülent Ecevit's Democratic Left Party (DSP), the center-right Motherland Party (ANAP), and the right-wing Nationalist Action Party (MHP), reflected this growing nationalist orientation. The DSP and MHP rode to power in 1999 on a tide of rising nationalism, fueled in part by the struggle against the PKK as well as Turkey's perceived rejection by Europe at the time. In many ways, their success was due to their ability to capture what Alan Makovsky has termed "Turkey's nationalist moment."[2]

Nationalist sentiment, along with closer measurement of Turkish national interest, has fueled a more active and assertive Turkish foreign policy. This has been reflected, in particular, in Turkish policy toward Syria, where in the fall of 1998 Turkey openly threatened to use military force to compel Damascus to cease its support for the PKK. It has also been reflected in the restrictive attitude that Turkey has taken toward American use of Incirlik air base. And it could be seen in the strong Turkish reaction to a proposed Armenian genocide resolution in the U.S. Congress in the Fall of 2000.

Another important trend has been the growing influence of the independent media, especially television. The media played an important role in the crisis over Imia/Kardak, which brought Turkey and Greece to the brink of war in January 1996, as well as in the first Chechnya crisis. Moreover, Islamist and other political groups now have their own television stations, giving them unprecedented access to a much broader cross section of the Turkish public. Public opinion and the media are now far more important factors in Turkish external policy than ever before.

At the same time, the growth of a dynamic private sector has served to weaken the role of the "strong state" and strengthen the power of civil society.[3] The business community, in particular, has emerged as an important political force in Turkey. The Turkish Industrialists' and Businessmen's Association (TUSIAD) and other institutions have emerged as outspoken advocates for reform and have proposed new policy initiatives on a broad range of social, economic, and political issues, including the Kurdish issue. Turkish entrepreneurs have also played a leading role in the expansion of economic ties to Russia, Central Asia, and the Middle East and have been at the forefront of the recent rapprochement with Greece. Together with the changing

---

[2]See Alan Makovsky, "Turkey's Nationalist Moment," *The Washington Quarterly*, Vol. 22, No. 4, Autumn 1999, pp. 159–166.

[3]See Henri J. Barkey, "The Struggles of a Strong State," *The Journal of International Affairs*, Vol. 54, No. 1, Fall 2000, pp. 87–105.

role of the military in Turkish society, the landscape for debate and policymaking on a range of issues, including foreign and security policy, is changing rapidly, with new actors operating alongside traditional elites.

Finally, the role of the military as a key actor in Turkish foreign and security policymaking continues to evolve in important ways, and more significant changes could lie ahead. On key issues, from the Kurdish issue in its regional context, to NATO policy, the military establishment has exercised a dominant influence. The military has also had a subtle, and sometimes not so subtle, role in the evolution of Turkish politics in recent years, although Turkey's tradition of direct military coups is probably a thing of the past. As Turkey faces stark choices regarding reform and European integration, the position of the military in Turkish society and policy has come under increasing criticism. Turkey cannot meet the EU's "Copenhagen criteria" without a substantial change in the role of the military. Moreover, it is doubtful that the Turkish military itself is monolithic in its opinion regarding important issues such as European integration, détente with Greece, and the path of reform in Turkey itself. Changes in the role of the military will be both an influence on and a product of a changing foreign policy.

WHY THIS BOOK?

In short, many of the traditional paradigms that characterized Turkey's international role and international relations over past decades are no longer valid. Turkey's domestic and foreign policies are evolving in new and important ways. At the same time, Turkey's geopolitical environment is changing, creating new opportunities but also new risks and vulnerabilities. Thus, a fresh look at Turkey's foreign and security policy is both timely and necessary.

This book explores these changes and their implications for Western policy. It expands on previous RAND work on Turkey.[4] However, it

---

[4]See, for example, Graham E. Fuller and Ian O. Lesser, *Turkey's New Geopolitics: From the Balkans to Western China*, Boulder, CO: Westview/RAND, 1993; Ian O. Lesser, *Bridge or Barrier? Turkey and the West After the Cold War*, Santa Monica, CA: RAND, 1992; and Zalmay Khalilzad, Ian O. Lesser, and F. Stephen Larrabee, *The Future of Turkish-Western Relations*, Santa Monica, CA: RAND, 2000.

is not a historical survey. Rather it focuses on key areas where Turkish policy is changing and examines the implications of these changes. Special attention is paid to security and defense issues because these issues are at the forefront of the Turkish debate, and because Turkish thinking on these matters will have important implications for where Turkey "fits" in the emerging Euro-Atlantic security architecture over the coming decades.

Chapter Two of this book examines the effect of domestic changes on Turkish foreign and security policy, taking into account the country's ongoing economic crisis. In particular, it focuses on the outlook for Turkey's traditionally "strong" state; pressures for and obstacles to modernization and reform; the rise of new policy actors in the private sector and elsewhere; the changing role of the military; the prospects for Islamism and nationalism; and the internal aspects of Turkish security perceptions. Finally, the analysis discusses the consequences of internal developments for Turkey's ability to play an active international role.

Chapter Three explores Turkey's relations with Europe in the aftermath of the EU's Helsinki summit. How will the decision to accept Turkey as a candidate for EU membership and subsequent developments affect Turkey's relations with Europe and with other key actors such as the United States and Russia? What does it mean for Turkey's domestic reform agenda? What problems lie ahead? How will the creation of a European Security and Defense Policy within the EU affect Turkey?

Chapter Four examines Turkey's relations with Greece and the Balkans. It focuses in particular on the prospects for a deepening of the recent détente between Ankara and Athens. It also assesses the broader implications of the rapprochement for relations with the EU, NATO and the United States. Finally, it looks at the implications of the Greek-Turkish relationship for stability in the Balkans and the Eastern Mediterranean. How strong is Turkey's new activism in the Balkans? What are the driving forces behind it? What are the implications for Turkey's relations with its key allies?

Chapter Five examines Turkey's relations in Central Asia and the Caucasus. Particular attention is focused on how Turkey's ties to the newly independent states in these regions will affect Ankara's relations with Russia and Iran. The chapter also examines "pipeline politics" and the prospects for energy cooperation in both regions. Finally, it analyzes the effect of domestic influences on Turkish policy toward Central Asia and the Caucasus.

Chapter Six examines Turkey's involvement in the Middle East. How are Turkey's relations with Iran, Iraq, and Syria likely to evolve? What are the implications of Turkey's deepening defense cooperation with Israel? How durable is it? What effect will uncertainty in the peace process have on relations with Israel and Turkey's wider regional interests? Finally, how are issues such as WMD, missile proliferation, and energy trends likely to shape Turkish foreign and security policy?

Chapter Seven focuses on Turkey's relations with the United States in light of evolving bilateral interests and policy concerns. In particular, it examines how new factors—the growing role of the EU, the Turkish-Israeli relationship, Turkey's increasing economic interaction with Russia, and Ankara's relations with other key regional actors such as Iran—are likely to affect Turkey's ties to the United States. Finally, it explores how domestic changes and new strategic concerns in the United States and Turkey may influence the evolution of bilateral ties.

The final chapter offers overall conclusions and observations and assesses the implications of a changing Turkish foreign and security policy scene for relations with the West. What problems are Turkey's evolution likely to pose for Western policy? Can we expect greater convergence or divergence in policy approaches? How can policy differences in areas of shared concern be reduced or eliminated? Finally, to what extent is Turkey's foreign and security policy evolution amenable to international, including Western, influence, and to what extent will Turkish external policy be driven by domestic forces?

Chapter Two

THE CHANGING DOMESTIC CONTEXT

Changes in the international environment are placing new pressures on Turkish policymakers and the Turkish public and are having important effects on Turkish policy. This is particularly true given the magnitude and rapidity of developments in adjacent regions, whether in the Balkans, the Caucasus, or the Middle East. These pressures alone would be stressful for Turkish foreign and security policymaking, which has a tradition of marked conservatism.

At the same time, Turkey confronts changes on the domestic scene that are arguably even more significant in their foreign and security policy implications. Turkey remains embroiled in a severe economic crisis that most Turks view as political at its base. The way Turkey responds to these economic and political challenges will shape Turkish society, perhaps for decades to come. It will also be a leading determinant of Turkey's freedom of action and the direction of Turkish policy on the international scene in the coming years.

This chapter explores key issues at the nexus of internal change and Turkey's foreign and security policy behavior, against a background of economic and political turmoil. These include the future of the state, the rise of new political and economic actors, the changing role of public opinion, the primacy of internal security considerations in Turkish policy, and the future role of Turkey's Islamists and nationalists.

AN ECONOMIC—AND POLITICAL—CRISIS

The financial crisis of November 2000, and the much more severe crisis of February 2001, were precipitated by a liquidity crisis exposing the fragility of a banking sector plagued by corruption and political favoritism. The proximate cause of the February crisis was a political clash between Prime Minister Ecevit and President Sezer, but the underlying causes are deep-seated and structural. The rescue package, organized by the IMF and the World Bank in the winter of 2000–2001, amounted to some $16 billion and was tied to a stringent program of austerity, privatization, and banking sector reform. By the fall of 2001, it became clear that further "bailouts" by international financial institutions would be required. The ongoing economic crisis has been especially painful for the more modern sectors and regions, with Istanbul particularly hard hit. The Turkish lira lost almost 50 percent of its value, virtually overnight, and nearly 65 percent by October 2001. Unemployment mounted rapidly in the financial sector, affecting many younger, urban Turks. The collapse of the commercial credit system had a pronounced effect on small and medium-sized enterprises across the country. Turkish and foreign observers began to openly speculate about the prospects for social unrest and more violent protest. The economic reform package, above all, changes in the banking sector, will have enormous implications for Turkish politics because the old system of patronage using state-controlled banks has been the basis for funding and influence of Turkey's leading political parties.

The foreign and security policy effects of Turkey's economic crisis could be profound and may be assessed at three levels. First, Turkey's near-term reform choices will shape the country's domestic development and foreign orientation. Since the end of the Cold War, observers have often described Turkey as being at a crossroads. In previous crises, Turkey has simply "muddled through" without pronounced changes in course. By contrast, the events of 2000–2001 have clearly led Turkey to a crossroads, by any definition, and muddling through is unlikely to suffice.

Turkey faces two possible paths. On the one hand, successful implementation of the IMF-inspired economic reforms will require a degree of political change that could facilitate more rapid reforms, greater democratization, and the steps required for closer integration in Europe and the continued modernization of Turkish society. This very positive path would encourage an active but Western-oriented and multilateral foreign policy—in short, a more moderate and predictable Turkey. On the other hand, Turkey's conservatism and statism (common on the left as well as the right) may impede economic and political reform, and a deepening crisis could strengthen already potent nationalist forces within the country. The result would be a more inward-looking Turkey, more sovereignty conscious, and more inclined toward a unilateral and less predictable policy on the international scene. In the worst case, a more chaotic and uncontrolled Turkey would have little energy and resources for foreign policy initiatives. A more unstable and less secure Turkey, or a Turkish collapse, would pose substantial dilemmas for both U.S. and European policy and would leave a series of regional vacuums from the Balkans to the Caucasus and the Middle East.

Second, the economic crisis has had a "reshuffling" effect on Turkish politics and society. Old cleavages between, for example, religion and secularism, have been replaced by new divides, principally over the question of political reform—support for the old order and the established parties, or something new. The debate over reform appears to cut across sectors, including business, the bureaucracy, and even the military, that have often been seen as monolithic. Reformers and conservatives are to be found at all levels and in all sectors. Turkey's secular establishment, long the leading interlocutor for the West, is being challenged by groups that have existed largely in the shadows and have traditionally had a less prominent role in the country's international engagement. The result of this struggle over leadership and reform is likely to have considerable influence on Turkish foreign policymaking in the future.

Third, the crisis is already having an effect on Turkey's key international relationships and the resources for national security policy. Notwithstanding its avowed distaste for international bailouts, the Bush administration has supported the IMF-led package of emergency assistance for Turkey. This support has not been linked to policy preferences in Washington, beyond insistence on adherence to the IMF reform plan. But requests for further support could well inspire a more explicit linkage to policy on Iraq, Iran, Cyprus, and other areas where American and Turkish views have differed. On the issue of counterterrorism, where American and Turkish policies are closely aligned, expanded cooperation is likely to strengthen the case for continued financial support to Ankara. Given Turkey's status as a candidate for EU membership, Washington might also insist that Turkey's EU partners bear a larger burden in any future assistance for Turkey. Europe, for its part, may see the economic crisis as justification for a very slow and cautious approach to the entire question of Turkish membership.

Turkey's regional position may also be affected. Economic stringency will undoubtedly complicate the military's ambitious modernization plans and may indirectly facilitate a reduction in both Turkish and Greek defense budgets. It could also slow the pace of defense-industrial cooperation with Israel, as well as procurement from the United States and elsewhere. Energy requirements have been a leading factor in Turkey's regional diplomacy. Access to energy is likely to remain an important objective in relations with Russia, Central Asia, Iran, and Iraq, but the rate of growth in Turkish energy demand may slow in a troubled economy. In Afghanistan, where Turkish forces have led international peacekeeping operations, financial stringency has made it even more imperative that Turkey's allies, principally the United States, subsidize the cost of Turkish participation.

Overall, Turkey's economic and political travails have interrupted an increasingly active Turkish debate about foreign and security policy and a more active set of external policies. Policymakers and the public are intensely focused on resolving the country's internal problems and on the search for new leadership. The conservative, evolutionary nature of Turkish foreign policy could be disturbed by these developments, with the potential for a more nationalistic or simply less energetic and more inward-looking approach.

A QUESTION OF VISION

Since the beginning of the Republic, ideology—in the benign sense of a guiding philosophy—has had a powerful influence over public policy in virtually all spheres. Atatürkism as an ideology may have lost some of its coherence and influence over the last decade, but the legacy of almost 80 years can still be felt strongly in many areas. In politics, Atatürkism has stood, above all, for secularism and the unitary character of the Turkish state. In economics, Atatürkism has stood for statism—a legacy that is only now being eroded under pressure of an economic crisis and the imperatives of reform in a "globalized" economy. In foreign policy, the Atatürkist tradition has emphasized nonintervention, a Western orientation, and vigilance with regard to national sovereignty.

In Atatürk's own vision, domestic and foreign policymaking were also closely linked. He championed the use of "the world power balance" and Turkish foreign policy to defend "the full independence and territorial integrity of the Republic." In negotiations surrounding the Lausanne conference in 1923, Atatürk asserted that the "foundation of foreign policy is a strong domestic policy, a strong domestic administration, and domestic organization. Domestic policy and foreign policy must always be linked."[1]

As Turkey enters the 21st century, these principles are being transformed by new actors and new issues in the Turkish policy debate. In the period since the Gulf War, in particular, Turkey has become a much more assertive actor in foreign and security policy. Ankara remains a relatively cautious player on the international scene and retains a strong preference for multilateral action in most areas, but nonintervention is no longer a meaningful description of Turkish policy. The country has at times flirted with more independent options. There is also a growing tension between Turkey's very traditional and strong sense of national sovereignty and the demands of integration as Turkey aims at eventual membership in the EU. Pressures from the IMF and other international institutions have also increased as a result of the country's economic troubles, and these too have spurred a nationalist reaction in some quarters. To the extent that Turkey progresses in its convergence with Europe, and as its society continues to modernize, this tension between sovereignty and integration is set to increase.

Viewed another way, change or the lack of change inside Turkey will be a critical determinant of Turkey's foreign and security policy options for the future. Without reform and democratization, not to mention sustained economic growth, the EU option is likely to be stressful, if not altogether foreclosed. The economic crisis of 2000-2002 has underscored this reality. Electoral triumphs by Turkey's Is-

---

[1]Quoted in Andrew Mango, "Reflections on the Atatürkist Origins of Turkish Foreign Policy and Domestic Linkages," in Alan Makovsky and Sabri Sayari, eds., *Turkey's New World: Changing Dynamics in Turkish Foreign Policy*, Washington, D.C.: The Washington Institute for Near East Policy, 2000, p. 10.

lamists, or nationalists, could recast Ankara's approach to Europe, Russia, and the United States. The persistence of Kurdish separatism would sustain a worldview, especially in Turkey's security establishment, that places internal security concerns at the top of the strategic agenda—a distortion of Turkish security planning that is at odds with the situation elsewhere within NATO. A focus on security problems inside Turkey is also likely to delay or impede movement toward Western patterns of civil-military relations—a potentially critical constraint on relations with Brussels and Washington.[2] The emergence of a much sharper discussion within Turkey over the future of its society, and whether to take risks with security and stability in pursuit of a more liberal system, have pushed staunch Kemalists within the Turkish establishment into a harder-line stance. As Cengiz Candar, a prominent journalist, has noted, "Kemalism is now a kind of state religion in its own right."[3]

Images of the Turkish internal scene have always had a pronounced effect on Western attitudes toward Turkey. This can be seen clearly in the European debate over Turkish EU membership, as well as in discussions in Washington on arms transfers, human rights, and the bilateral relationship as a whole. This tradition of analyzing Turkey and Turkish policy "from the inside out" is a very old one. In the Ottoman period, the nature of Turkey as a society was a key element shaping European perceptions. The five-hundred-year competition between the Ottoman empire and the West was not just a geopolitical competition but a competition between societies and, above all, religions. In the waning years of the empire, Western perceptions focused on Ottoman "backwardness" and these, in turn, influenced opinion in support of national independence movements in the Balkans and assessments of Turkey's ability to help contain Russian ambitions in the Eastern Mediterranean. The potential for developments inside Turkey to affect surrounding regions and the interests of extraregional powers is a central tenet of much recent analysis of

---

[2]For a highly critical assessment of the role of the Turkish military as an impediment to Turkish liberalization and as an obstacle to Turkey's European ambitions, see Eric Rouleau, "Turkey's Dream of Democracy," *Foreign Affairs*, Vol. 79, No. 6, November/December 2000, pp. 100–114.

[3]Cengiz Candar, "Atatürk's Ambiguous Legacy," *The Wilson Quarterly*, Vol. XXIV, No. 4, Autumn 2000, p. 95.

Turkey as a "pivotal state."[4] It is arguably an important factor in the Western calculus regarding assistance to Turkey in its economic difficulties and encourages many Turks to believe that what happens inside Turkey is simply too important for the West to ignore.

WHITHER THE STRONG STATE?

Turkey is an example of a "strong state," that is, a society in which the state is at the center of public policymaking and the notion of state sovereignty is highly developed and unalloyed.[5] The idea of the strong state does not necessarily imply a powerful or capable state in practical terms (although, overall, the Turkish state has played this role in some areas) but rather a pervasive, doctrinal attachment to the primacy of the state. Republican Turkey emerged in 1923 at a time of authoritarian statism in postrevolutionary societies across Europe and Asia. Indeed, parallels have often been drawn between the emergence, in roughly the same period, of strong, centrally directed states bent on modernization in both Turkey and the Soviet Union. For almost 80 years, the Turkish state has had a pervasive role in virtually all aspects of Turkish life. The model of Turkey as a secular, Western-oriented society was promoted—very successfully—from the top down, from the earliest years of the Republic. Economic policy was shaped from the center on a statist pattern, with high levels of government ownership and oversight.

The idea of national sovereignty that guided the formation of the Republic and that persisted essentially unchallenged until the 1990s was based largely on 19th century European ideas.[6] In important respects, the orientation of the early Republic was itself a reaction to threats to Turkish sovereignty and territory at the close of the Ot-

---

[4]See Alan Makovsky's chapter on Turkey in Robert Chace et al., eds., *The Pivotal States*, New York: Norton, 1999, pp. 88–119.

[5]See Henri J. Barkey, "The Struggles of a Strong State," pp. 87–105; and the discussion of the Turkish case in Ian O. Lesser, *Strong States, Difficult Choices: Mediterranean Perspectives on Integration and Globalization*, Washington: National Intelligence Council/RAND, 2001.

[6]See the discussion of Turkish nationalism in Ernst Gellner, *Encounters with Nationalism*, Oxford: Blackwell, 1994, pp. 81–91; and Anthony D. Smith, *National Identity*, Reno: University of Nevada Press, 1991. See also Hugh Poulton, *Top Hat, Grey Wolf and Crescent*, New York: NYU Press, 1997.

toman period and shortly thereafter, including the restrictive provisions of the Treaty of Sèvres (1920), the loss of Ottoman territories in the Balkans and the Middle East, and the Greek military intervention in Anatolia. Out of these experiences came a consolidated and fiercely defended idea of a unitary Turkish state, with state-directed modernization and Westernization as key elements, and with a very cautious approach to international affairs. Atatürk's famous dictum "peace at home, peace abroad" captured the spirit of this period in which Turkey sought to reduce its exposure to further Western intervention (again, there are notable parallels to the Soviet concept of building "socialism in one country"). The experience of this period has left an enduring legacy of suspicion, among the Turkish public and elites, regarding Western aims in Turkey and its region. European and American observers may find Turkish fears of a Western-inspired breakup of the Turkish state unreasonable, but the power of these images cannot be discounted in Turkey, even today. Turkish views on the Kurdish problem and the situation in Northern Iraq, as well as reactions to American Congressional debate on an Armenian genocide resolution, point to the enduring nature of these suspicions.[7] Not a few Turks will even suggest that Western "conditionality" in support for economic assistance to Turkey is rooted in a desire to shape and perhaps limit Turkey's regional role.

Today, the greatest challenges to the traditional role of the Turkish state are internal, but these challenges are reinforced by pressures from outside, from the requirements of integration with Europe, regardless of Turkey's membership prospects, to the effects of globalization on Turkey's society and economy. Arguably, the future of the strong state is the central question for Turkey's evolution in the coming years.

The primacy of the Turkish state is being challenged from many directions. First, and most fundamentally, the rapid if sometimes erratic growth of the Turkish economy since the economic liberalization of the Özal years has made the task of economic management

---

[7]A nonbinding resolution concerning the Armenian genocide was introduced in the U.S. Congress in October 2000 but later withdrawn. Introduction of the resolution touched off a vigorous, nationalist reaction in Turkey. Many Turks saw the resolution concerning events in the last years of the Ottoman empire as an affront to modern Turkey's national honor.

far more complex, and the central government in Ankara no longer enjoys a monopoly in economic affairs.[8] Business associations, individual entrepreneurs, unions, and Turkey's increasingly active media all want a say on economic matters. Although successive Turkish governments have been committed to the idea of privatization, progress in this area has been slow and characterized by a lack of transparency. It is notable that the Islamist Refah Party, once in government, was among the most adept at using the privatization agenda to build a constituency of supporters in the private sector.

The rise of a very dynamic Turkish private sector has, however, created an important counterweight to the sluggish and inefficient state enterprises that still make up a large proportion of the Turkish economy. This process has brought to the fore a number of prominent, family-controlled business empires, but it has also led to the rise of a large number of small and medium-sized enterprises, many led by prosperous Anatolian families that had traditionally been involved in agriculture. The net result has been the emergence of a very diverse private sector, with a range of perceptions about politics and foreign affairs. These range from the secular, reformist, and internationally minded members of TUSIAD to MUSIAD (the Independent Turkish Businessmen's Association), an active group with a traditional and religious orientation. These very diverse elements are nonetheless generally united on the desire to reduce the role of the state in Turkish society.[9]

The coexistence of a dynamic private sector alongside a large public sector also means that many private enterprises, including those in the financial sector, rely heavily on their relationship to the state.[10] This phenomenon is, in part, responsible for the rise of corruption in Turkey as a growing public policy problem and represents yet an-

---

[8]Over the last fifteen years, Turkey has enjoyed some of the highest rates of growth in the Organization for Economic Cooperation and Development (OECD). Between 1995 and 1998, growth averaged 8.5 percent. Since that period, growth has been slower, partly because of the costs of recovery from the 1999 earthquake, estimated at $5.7 billion. Growth in 1999 was 6.4 percent. *Turkish Economy 1999–2000*, Istanbul: TUSIAD, July 2000.

[9]See Stephen Kinzer, "Businesses Pressing a Reluctant Turkey on Democracy Issues," *New York Times*, March 23, 1997.

[10]See Aysa Bugra, *State and Business in Modern Turkey: A Comparative Study*, Albany: State University of New York Press, 1994.

other angle of criticism toward the role of the state. The links between the state, banks, and large-scale business in Turkey have also been at the center of allegations of "crony capitalism" as Turkey experiences its worst economic crisis since World War II. The most successful political movements in recent years, including the Islamists and National Action Party, have capitalized on public cynicism about Turkey's aging political class and have taken a strong stance against cronyism, corruption, and the lack of transparency.[11] Perhaps more serious, over the last decade Turkey has seen a general expansion of the illegal sector, above all drug trafficking and money laundering, along with more exotic problems such as the smuggling of people and even nuclear materials. These activities exist largely outside the bounds of the state, but some of the most prominent scandals of recent years have focused on alleged links between organized crime, terrorism, and the state. The unstable situation in Southeastern Anatolia, with its own war economy surrounding the battle with the PKK, has been a breeding ground for the Turkish illegal sector.

Second, Turkey's demographic growth and, more significantly, the tremendous movement of population from the countryside to the major cities over the past decades have placed great strains on the state's ability to provide essential services and to provide for social welfare. In 1945, 25 percent of the Turkish population lived in urban areas. By 1997, this figure was 65 percent, and the percentage continues to increase.[12] (There is some anecdotal evidence that the economic crisis is causing a slight countermovement from Istanbul back to the countryside as employment prospects in the city have worsened.) Turkey's population stands at roughly 67.8 million today, and even with recent reductions in the rate of population growth, Turkey's population is likely to reach 100 million early in the 21st century.

---

[11]See Ziya Onis, "Neo-Liberal Globalization and the Democracy Paradox: The Turkish General Elections of 1999, *The Journal of International Affairs*, Vol. 54, No. 1, Fall 2000, p. 306.

[12]*Turkey's Window of Opportunity: The Demographic Transition Process and Its Consequences*, Istanbul: TUSIAD, 1999, p. 21. See also Andrew Mango, *Turkey: The Challenge of a New Role*, Westport: Praeger, 1994, pp. 64–75.

At the same time, the modernization of Turkish society and exposure to Western patterns have raised public expectations about what the state should provide. Turkey's Islamist movements, Refah (Welfare) its successor, Fazilet (Virtue)—both legally banned—and newer Islamic groups, such as the Justice and Development Party, have capitalized on this growing gap between public expectations and the ineffectiveness of state institutions. This is one explanation for the consistent success of Turkey's Islamists in municipal elections across the country. Dissatisfaction with the role of the state was dramatically demonstrated in the wake of the major earthquake that struck Turkey in August 1999. Under very stressful conditions, the state appeared incapable of effective civil emergency management. The earthquake may not have spelled the end of the strong state, as many predicted at the time, but it has certainly left an enduring legacy of dissatisfaction. To a degree, this dissatisfaction with the state, and the tendency at many levels of society to organize affairs, to the extent possible, without reference to the state, is a common phenomenon across the Mediterranean.[13]

A third challenge to the role of the Turkish state arises from external pressures. Simply put, the current role of the state as enshrined in the Turkish Constitution is incompatible with the objective of closer integration in Europe and an obstacle to meeting the criteria for the opening of accession negotiations with the EU. The incompatibility is clear at the level of economic policy, where convergence would require a substantial reduction in the role of the state even to meet European norms. More fundamentally, integration implies a substantial diminution in state sovereignty as policy and administration in key areas are subordinated to European procedures.[14] Movement toward a more dilute and modern notion of sovereignty is at the heart of much Turkish ambivalence about the implications of membership, but the pressures for modernization in this sphere are very strong. If Turkey's European aspirations are not frustrated, and the

---

[13]This has sometimes been described as the "Italian model," in which an efficient private sector coexists alongside ineffective state institutions, and individuals seek private arrangements for the provision of essential services.

[14]See David Barchard, *Building a Partnership: Turkey and the European Union,* Istanbul: Turkish Economic and Social Studies Foundation, 2000.

post-Helsinki path remains open, significant changes in the role of the state can be anticipated, affecting many aspects of Turkish life.

The question of the role of the state is also central to the outlook for political reform, civil-military relations, and human rights. Turkey is a functioning if very imperfect democracy, and the process of democratization in Turkey is well advanced by the standards of adjacent regions. There is an extremely active public debate in Turkey over questions of democratization and reform—a debate that has acquired even more vigor and urgency in the context of pressing economic problems. Nonetheless, there are persistent problems concerning human rights, the status of minorities, and the governance of regions within Turkey that also turn on the role of the state. The tradition of strong central authority and concern about threats to the unitary character of the state (read Kurdish separatism) have made the discussion of decentralization, and especially forms of regional autonomy, anathema for Ankara. The question of Kurdish rights, against the background of a violent insurgency and counterinsurgency campaign in Southeastern Anatolia, has made this a highly charged issue for Turks, and for Turkey's partners in the West. By most accounts, the human rights situation in Turkey has improved in some respects but is not improving rapidly enough to satisfy Western opinion—or to satisfy many Turks.[15] There is now greater transparency with regard to human rights abuses, and the waning of the PKK insurgency has created a better climate in the towns and cities of the Southeast. But abuses persist, and political rights (e.g., the treatment of "thought crimes") are limited in ways that continue to surprise Western observers.

Many Turks are hopeful that the progressive modernization of Turkish society, efforts at constitutional reform, and ever-increasing transparency will eventually change those aspects of Turkish political culture that have fostered human rights abuses. The advent of a reform-minded president, Ahmed Necdet Sezer, was widely interpreted as a positive sign. Sezer has, for example, taken the position (opposed by Turkey's military as well as by Prime Minister Ecevit) that state "decrees with the force of law" may not be used in lieu of

---

[15]See U.S. Department of State, *Turkey: Country Report on Human Rights Practices for 1998*, Washington, D.C., 1999. This qualified, critical tone has continued through subsequent yearly reports, including the report for 2001.

parliamentary measures to remove alleged antisecularists and separatists from Turkey's state bureaucracies. Under pressure from international financial institutions, Turkey is now struggling to implement a series of economic and political reforms that will also require constitutional changes. If successful—and the outlook is unclear—the ground may be paved for more far-reaching changes in democratization and human rights.

THE ROLE OF THE MILITARY

The role played by the Turkish military has been changing in important ways. Historically the armed forces have occupied a privileged position both in the Ottoman period and in the Turkish Republic. In the late Ottoman period and in the early days of the Turkish Republic, the military spearheaded the modernization process. Six out of ten presidents of the Turkish Republic—including Kemal Atatürk, the founder of the Turkish Republic—were high-ranking military officers. Today many Turks worry that the Turkish military's doctrinaire interpretation of Atatürk's policies is becoming an obstacle to the democratization and modernization of Turkish society. But understanding military attitudes in Turkey is not an easy task, and opinion within the military establishment may be more dynamic and less monolithic than is usually imagined.[16] In any event, the military's stance on reform, including the reform of civil-military relations, will be a key factor shaping the future of Turkish society and its international relations.

The Turkish military has acted as custodian of the Kemalist legacy, seeing its mission as not only to defend the territorial integrity of the Turkish state against external threats but also to protect it against internal challenges. The armed forces have intervened in Turkish politics four times in the postwar period when they felt that the Kemalist legacy was under threat. The most recent instance was in 1997 when the military forced the ouster of Prime Minister Necmet-

---

[16]Turkish observers have pointed to the growing differences of view within the military. The respected Turkish journalist Mehmet Ali Birand, for example, has noted; "We can not speak anymore of a Turkish Armed Forces that has an uniform view, adopting as its own view the ideas expressed by its commander. There are different views within the ranks of the military as well." See Mehmet Ali Birand, "Why does the military keep silent?" *Turkish Daily News*, July 17, 2002.

tin Erbakan, head of the Refah Party, in what was widely interpreted by Turkish and foreign media as a "silent coup."

The military's power is institutionalized through a variety of organizations. The most important of these is the National Security Council (NSC). Although technically the NSC makes only "recommendations" to the Council of Ministers, its recommendations can be tantamount to orders—as Prime Minister Erbakan was reminded in 1997. When Erbakan attempted to circumvent the military's 20 recommendations on curtailing Islamist activity by sending them to the parliament, rather than carrying them out, the military—backed by Turkey's wider secular establishment—forced his eventual resignation.

The relationship between the military and Turkey's political leadership has often been uncomfortable, even on foreign policy matters. Özal pressed for active Turkish participation in the Gulf War coalition, against the advice of a more cautious military establishment, eventually leading to the resignation of Chief of Staff General Torumtay. Özal also took a softer line toward Islamism and favored a more liberal policy on the Kurdish issue. Since Özal's death in 1993, the military has gradually won back the power lost during his tenure, bolstered by success in the counterinsurgency campaign against the PKK and a continuing, central role in opposition to religion in political life.

Nonetheless, the post-Helsinki dynamics in Turkey's international relations and the ambivalent position of the military on the reforms necessary for Turkish convergence with Europe, suggest that Turkey's military may find it difficult or unappealing to maintain its traditional role in the coming years. The military has a clear stake in overcoming Turkey's current financial crisis, for the sake of keeping defense modernization plans on track, and because the military is also a large investor and participant in the Turkish economy.

The extent of reforms in the area of governance and human rights will have a profound influence on the character of Turkey's relations with the West and above all with the EU. Indeed, changes in these areas are an essential prerequisite for progress toward accession negotiations. Civil-military relations are an important part of this equation. Analyses of the Turkish scene have traditionally emphasized the role of the military in Turkish governance. With the experience of three military *coups d'état* in the postwar period, and the intervention of the military behind the scenes to engineer the end of an Islamist-led coalition in 1997, the military remains an influential actor on the political scene. In the areas of key concern to the military—secularism and the unitary character of the state (i.e., separatist threats)—their influence has not been exercised in a vacuum. Many in Turkey's secular elite, and those members of a much wider group for whom the military retains high prestige and credibility, quietly support the traditional model of civil-military relations. This may change, however, as those with a stake in more rapid reform and convergence with Europe acquire greater influence in Turkish policymaking.

Many Turks today would probably agree that the role of the military in Turkish society and politics is changing and is likely to change further under the pressures of modernization and with the emergence of competing policy elites (discussed below). Here, too, the future of the relationship with Europe is likely to be a strong influence, as the current pattern of civil-military relations in Turkey would not meet the EU's requirements for membership.[17] By this measure, it is even uncertain if Turkey would be considered for NATO membership if it were applying today. The survival of what may appear to be an anachronistic military role in Turkey may be explained by the transitional nature of the Turkish situation, in particular, the special role of the military as guardians of the secular state and the Atatürkist tradition. It may also be explained by the primacy of security issues, many of which are internal, in Turkish perceptions and policies. Finally, and not least, the military has been an effective institution, perhaps the most effective institution within the Turkish state. The military establishment is seen as largely untainted by corruption, despite being a very important economic actor in its own right. This is a function of the sheer size of the military establishment—the second largest in NATO after the United States—its large claim on state

---

[17]The role of the Turkish military has become a flashpoint for European criticism of Ankara and pessimism regarding the country's EU prospects. See, for example, the analysis by Eric Rouleau, "Turkey's Dream of Democracy," *Foreign Affairs*, Vol. 79, No. 6, November/December 2000, pp. 102–114.

spending, and its pension, foundation, and commercial holdings.[18] Retired senior officers are commonplace on the boards of Turkey's large holding companies.

It is also worth noting that opinion within the military establishment is not monolithic, even on controversial questions such as strategy toward Turkey's Islamists or policy toward the EU. Some Turkish observers are beginning to suggest that there is now a debate within the military itself about the future role of their institution in Turkish society and the likely need for changes in civil-military relations. The extent of these changes is sure to be a leading determinant of Turkey's integration and reform prospects over the next decade.

NEW ACTORS IN A CHANGING SOCIETY

The military as an institution has been important, not only in direct policy terms but also as a vehicle for upward mobility, education, and training in Turkish society. Since the early years of the Republic, the military has been extraordinarily successful as a vehicle for socialization across the country. For decades, young Turks in the provinces aspired to military careers and the military as an institution enjoyed high prestige among average Turks.[19] The military ran many of the best and most modern educational institutions in the country. Although the role of the military, especially the Turkish General Staff, in Turkey's domestic politics (e.g., its influence on Turkey's coalition arrangements, education policy, and the headscarf question) has been the most controversial issue from a Western perspective, and from the perspective of Turkish liberals, the dominant position of the military in foreign and security policy is, perhaps, even more complete.

With few exceptions, Turkey has not had a well-developed cadre of foreign and security policy experts outside government circles. Ex-

---

[18]Defense spending was 5.2 percent of GDP in 2000. The involvement of the military in a variety of non-defense projects suggests that this figure hardly reflects the full extent of military engagement in the economy. *IISS Military Balance, 2000–2001,* London: IISS, 2001.

[19]Turkey's military culture, as it has evolved over recent decades, was described in considerable detail in Mehmet Ali Birand's controversial book, *Shirts of Steel: An Anatomy of the Turkish Armed Forces,* London: I. B. Tauris, 1991.

ternal and defense policy has traditionally been almost the exclusive preserve of professional diplomats and, not least, high-level military officers. The military's capacity for analysis of these issues has actually been bolstered in recent years by the expansion of the staff of the National Security Council. One consequence of the military's extraordinary role has been a marked asymmetry in Turkey's dialogue with allies on regional and defense matters. Whereas Western governments generally view their high civilian officials as the key interlocutors on foreign and security policy, the corresponding civilian officials in the Turkish Ministry of Defense are, in practical terms, subordinate to the leadership of the Turkish General Staff. Thus, Turkey's model of civil-military relations not only affects internal politics, but has shaped the country's dialogue with Europe and the U.S. on a range of issues. This is almost certainly one of the factors responsible for the continued predominance of security issues in Turkish-American relations.

Some of the elements pointing toward change in the role of the military and civil-military relations have already been suggested. Another element of equal or perhaps greater importance concerns the rise of other actors and "power centers" outside the military and outside the state. The transformation of the Turkish economy and the dynamism of the Turkish private sector have become a leading vehicle for change in Turkish society. Young Turks who might once have aspired to careers as military officers or as state employees are now more likely to seek careers in the business world—although this too could change if the country's economic problems persist. Moreover, Turkey's famous state universities and military schools are now challenged by a host of private colleges and universities, many with impressive funding from private entrepreneurs and foundations. At the primary and secondary level, the network of religious schools (the "Imam-Hatip" schools) expanded dramatically in the 1990s, a trend paralleling the rise of Islamic politics at the local and national levels.[20] Recent legislation has placed limits on the scope of the religious schools. But they have played a role in educating a generation

---

[20]The Imam-Hatip schools are high-school-level *lycées* with a religious orientation. Their graduates may go on to private universities, but not, for example, military academies.

of students who are now to be found in Turkey's bureaucracies and elsewhere.

One consequence of this expansion of alternative education and career paths has been the emergence of a generation with new views about Turkey and Turkey's international role. The information revolution has clearly played a part, as well. Even provincial towns in Central Anatolia now sport Internet cafes, and the wide availability of satellite television and the proliferation of private stations have encouraged a far broader world view. The net effect of these changes on Turkish opinion is, however, an open question. These trends have certainly contributed to the erosion of the Atatürkist tradition in its various manifestations. But they have also sharpened the political debate, and strengthened support for more extreme views, especially among the young. In this context, it is noteworthy that in the 1999 general elections, younger voters formed the basis of National Action Party success.[21]

The vigor of the Turkish private sector over the past decade has had important consequences for public debate and policymaking, with meaning for the country's international relationships. Turkey's large holding companies (and a much larger group of small and medium-sized enterprises) have recognized and have begun to articulate their policy interests across a range of issues. The traditional reliance on personal relationships as a means of access and influence with state officials has been augmented by the emergence of private sector institutions devoted to discussion, analysis, and ultimately lobbying on questions of concern. Such institutions, including business associations and a small number of independent "think tanks," are commonplace in the West but are a very recent phenomenon in Turkey.[22] They are beginning to play a role in articulating the policy

---

[21]MHP's youth wing has attracted attention for its extremism and occasional violence, harkening back to the behavior of MHP in the 1970s and 1980s when the party was linked to right-wing terrorism in Turkey.

[22]TUSIAD, in particular, has emerged as a key source of analysis and policy commentary on the Turkish scene and Turkey's foreign relations. It has taken controversial, reformist positions on a number of issues, including the situation in the Southeast. TESEV (the Turkish Economic and Social Studies Foundation) in Istanbul was established in 1961 but is now becoming more active and influential, especially on Turkish-EU relations. The newly formed Istanbul Policy Center at Sabanci University is another policy-oriented institution to watch.

interests of particular circles, although the longer-term outlook for these institutions will be heavily dependent on the health of the Turkish economy and contributions from private sector sponsors. Notably, these institutions are among the most interested in building international ties and tend toward a liberal, reformist outlook. As a group, they represent an increasingly important set of interlocutors for Western officials and unofficial observers of the Turkish scene and encourage a more activist but multilateral approach to Turkey's foreign policy interests.

PUBLIC OPINION AND THE MEDIA

In the period since the Gulf War, public opinion has emerged as an increasingly important factor in Turkish foreign policy, spurred by the expansion of the private media and its growing role as a shaper of opinion. Some aspects of this phenomenon, including the concentration of media assets in the hands of a few large conglomerates, are controversial and the subject of considerable debate in Turkey (as they are elsewhere).[23] With the concentration of media ownership and the tendency for media holding companies to diversify their activities, including involvement in the financial sector, Turkey's economic crisis has contributed to the turmoil in Turkish journalism. Organizations such as CNN-Turk and other major stations, as well as newspapers, have recently seen large-scale layoffs after years of rapid expansion. The Dogan group currently dominates Turkish media ownership, and some Turkish observers point to the possible emergence of a "Turkish Berlusconi" in the country's tumultuous politics.

At the same time, the proliferation of media outlets, especially television, has brought a more diverse set of voices to the Turkish debate. In some cases, notably television stations with a religious orientation, the non-mainstream media have faced official and unofficial pressures over the nature of their programming, including temporary bans. But stations such as the religiously inclined "Channel 7" have nonetheless survived on a profitable basis. Papers such as *Yeni Safak* have attracted a sophisticated secular reader-

---

[23]See, for example, the discussion in Andrew Finkel, "Who Guards the Turkish Press? A Perspective on Press Corruption in Turkey," *The Journal of International Affairs,* Vol. 54, No. 1, Fall 2000, pp. 147–166.

ship—and some respected secular journalists—despite their religious orientation, because they are seen by many readers as objective and independent.

A more active and diverse media have been a vehicle for the rise of public opinion as a factor in the traditionally closed world of Turkish external policymaking. Several elements are worth noting in this regard. First, the engagement of public opinion on foreign policy issues has been encouraged by developments in Turkey's region over the past decade. The Gulf War marked a turning point in this respect, as the events in Northern Iraq first reinforced and then complicated Özal's presentation of his strongly pro-Western stance. The promise of more-rapid Turkish progress toward EU candidacy and other prospective benefits allowed Özal to successfully present the case for Turkish action, including the use of Turkish bases for the strategic bombardment of Iraq, and the closure of Iraqi oil exports through Turkey by road and pipeline. Public opinion in this case was an important element in Özal's ability to pursue an activist policy against the inclinations of the Turkish military. In the wake of the war, however, public opinion became critical of the costs of participation in the coalition and of Turkey's failure to receive sufficient compensation from the West. Many Turks came to regard, and still regard, the Gulf War as a catalyst for Kurdish separatism and a continuing cost to Turkey in the form of lost revenue from trade with Iraq and pipeline fees. Public opinion remains a major constraint in Ankara's policymaking toward Northern Iraq, including renewal of Operation Northern Watch.

Events in Bosnia also galvanized Turkish opinion and encouraged a more active, albeit multilateral, stance from Ankara. Turks have sympathized with the plight of the Bosnian Muslims and feel considerable affinity for Muslim communities elsewhere in the Balkans. Turkish affinities have also been aroused by Azerbaijan's dispute with Armenia over Nagorno-Karabakh, by successive conflicts in Chechnya, and by the crisis in Kosovo.[24] These crises also took

---

[24]It is interesting to note that the public reaction to the Kosovo crisis has been notably less animated than in the case of Bosnia. This may be explained by mixed attitudes toward the Muslim Kosovars, who have not always been on good terms with the Turkish minority in the region. Moreover, Turks, while sympathetic to the plight of the

on a strongly nationalistic and at times anti-Western (mainly anti-European) flavor in the face of perceived Western inaction. Periodic violence by Chechen sympathizers, including the April 2001 hostage-taking at the Swisshotel in Istanbul, seems designed to play to media attention and, ultimately, public opinion.

Second, the last decade has seen the rise of distinct "lobbies" within Turkish society. This was particularly observable in relation to the Bosnian and Nagorno-Karabakh conflicts and, of course, with regard to Cyprus. Crises in the Balkans, the Caucasus and the Eastern Mediterranean have come at a time when many Turks are rediscovering their regional roots. Bosnian and Azeri Turks, in particular, are to be found in some numbers among Turkey's elite, and their voices have been influential in shaping attitudes in Ankara. Similar ethnic identifications and attempts at policy influence—or at least symbolic gestures—have occurred on behalf of Bulgarian Turks, Tatars in Russia and Ukraine, the Gagauz in Romania, and the Uighur Turks in Western China. These associations are now very much a part of the Turkish scene.[25] In a very real sense, developments around Turkey's borders have stimulated more discussion inside Turkey about what it means to be a Turk and have encouraged tentative debate about Turkey as a multicultural society. This trend has also had an influence on the definition of Turkey's international interests, which clearly extend to include the status and treatment of Turks abroad, particularly those in Germany. The growing role of public opinion as a factor in Turkish policymaking has also been associated with a rise in nationalist sentiment, especially in circumstances, as in Bosnia, where Turkey's Western partners appeared inactive or indifferent.

Third, the public opinion factor complicates Turkish crisis management and has produced considerable unease among Ankara's foreign and security policy functionaries who are unused to this reality. In the Imia/Kardak crisis of 1996 between Greece and Turkey, for example, events were driven (on both sides) by aggressive television journalists and responsive public opinion, to the extent that Turkish diplomats and officers worried about losing control of the situation.

Kosovars, are generally opposed to new separatist movements (echoes of the Kurdish problem) and the alteration of borders in the Balkans.

[25]See Andrew Mango, "Reflections on the Atatürkist Origins of Turkish Foreign Policy and Domestic Linkages," pp. 9–19.

Similarly, in the hijacking of a Turkish Black Sea ferry by Chechen extremists in 1996, negotiations with the hijackers were preempted with the arrival, by helicopter, of Turkish journalists who assumed the role of negotiators, with the Turkish public held in rapt attention. Incidents of this type suggest a very different domestic environment for crisis management than has traditionally prevailed in Turkey.

SECURITY POLICY THROUGH A DOMESTIC LENS

In common with many other states around the Southern Mediterranean, Turkish debate and policy on security matters are strongly influenced by internal security concerns. This outlook is somewhat outside the NATO mainstream but is not completely alien to the Alliance, as attested to by Spain's preoccupation with Basque terrorism, Britain's struggle against terrorism in Northern Ireland, France's concern about spillovers of North African extremism, and the growing interest in "homeland defense" spurred by the recent experience of disastrous terrorist attacks in the United States. Yet, in the Turkish setting, questions of internal security are central to the political landscape and the worldview of Turkey's traditional foreign and security policymakers. Increasingly, problems of energy supply and non-traditional challenges in the areas of crime, refugees, and the environment are also being seen as part of the security agenda, with strong domestic linkages.

As noted above, Turkey's military establishment continues to have a central role in defining and vetting the policy agenda, and this establishment has been extraordinarily consistent over the past decade in describing the fight against Kurdish separatism and the fight against Islamism as the leading security challenges for the Turkish state. The struggle against Islamism has, by and large, been carried out through political and legal means and has resulted in the termination of Turkey's experiment with an Islamist-led government, the banning of Refah and its successor the Virtue Party, and the prosecution of leading figures in religious politics, including Recip Tayip Erdogan, the former mayor of Istanbul, and former Prime Minister Necmettin Erbakan. The political ban on Erdogan was lifted in July 2001, and he emerged as a very popular leader of the religious/reformist wing of the now defunct Virtue Party in national politics.[26] His political future as head of the new Justice and Development Party remains unresolved, however, in light of ongoing legal challenges.

For all the effort devoted to the suppression of Islamic politics, a substantial undercurrent of religion has survived and may be poised to emerge again in a more reform-minded vein. Turkey's extensive network of religious orders (*tarikat*) has operated underground since the time of Atatürk but reportedly retains substantial and influential memberships. Center-right parties such as True Path (DYP) and ANAP have always had their religious wings, and with the large number of deputies who entered parliament under the Virtue Party banner, the bloc of religious conservatives is very large. In this context, the controversy over the wearing of headscarves in government institutions, including schools, has been a political as well as a cultural dispute, pitting a more overtly religious public against the tradition of state involvement in all aspects of Turkish life.

The struggle against Islamic groups acquired a more direct security dimension with the crackdown on Turkey's violent and shadowy Hizbullah network (unrelated to the Lebanese movement of the same name). This network is alleged to be responsible for numerous assassinations and disappearances over the past decade, including a number of high-profile terrorist incidents. Turkish and foreign observers also link Turkish Hizbullah to organizations tolerated and perhaps supported by Ankara in the past as part of the counterinsurgency campaign against the PKK. 1999 estimates suggested that Hizbullah may have as many as 25,000 adherents, including 4,000 armed militants.[27] Revelations about these groups and the security operations against them have given a harder edge to the question of Islamism in Turkey at a time when more mainstream Islamic movements such as Erdogan's reformists within the Justice and Development Party, the Felicity Party and the Fethullah Gülen group are taking pains to appear centrist. Against this background, the attitude

---

[26]Recent polls suggest that Erdogan is Turkey's most popular politician. Douglas Frantz, "Turkey's Leaders Uneasily Await a New Party," *New York Times*, August 3, 2001, p. 10.

[27]"Turkey's Divided Islamists," *IISS Strategic Comments*, Vol. 6, Issue 3, April 2000.

of the military and secular elites to the Islamist question shows few signs of weakening. Although Europe has little sympathy for Islamists, the EU factor could constrain Ankara's ability to bar Turkey's Islamic parties from political life—a fact reflected in support for Turkish accession among Turkey's mainstream Islamists. In the meantime, there is growing concern that armed Islamic groups on the margins of Turkish society may take over the mantle of extremist opposition to the state, filling a vacuum left by the waning of the PKK challenge.[28]

The second overarching internal security challenge has been Kurdish separatism. Even before the capture of PKK leader Abdullah Öcalan, Turkish security forces had essentially defeated, but not eliminated, the PKK insurgency. The struggle against the PKK has claimed perhaps 30,000–40,000 lives on all sides since the mid-1980s. This little-reported war inside a NATO country imposed immense costs on the Turkish economy and society and distorted the Turkish political scene in fundamental ways. The war with its terrible human rights abuses—inflicted by both sides—continues to damage Turkey's international standing even as the struggle in the Southeast winds down.

The defeat of the PKK insurgency can be ascribed to a number of factors. The PKK campaign inside Turkey relied heavily on access to bases in Northern Iraq and Syria (and in Lebanon's Bekaa Valley) and practical support from Damascus. Since the mid-1990s, Ankara has pursued a cross-border strategy in the fight against the PKK, deploying its security forces over the Iraqi border to disrupt PKK operations and to fight the war, to the extent possible, on Iraqi rather than Turkish territory. The result has been the establishment of a *de facto* Turkish security zone in Northern Iraq, comparable in some respects, although by no means all, to the zone held until 2000 by Israel in Lebanon. The capture of PKK leader Abdullah Öcalan and the end of Syrian support as a result of a credible threat of Turkish intervention severely weakened the PKK's capacity for operations inside Turkey. This, in turn, touched off a struggle over strategy and leadership within the PKK and Kurdish organizations based in Europe, further weakening the PKK position. Even before the capture of

---

[28]Ibid.

Öcalan, Turkish forces had considerable success in containing PKK operations, largely as a result of improvements in training, tactics, and equipment.

The security situation in the Southeast has improved markedly, especially in cities such as Diyarbakir, and it is unlikely that the PKK will be able to reconstitute itself to carry out operations on the scale of the early 1990s. But perhaps 5,000 to 10,000 PKK militants remain in Turkey and Northern Iraq and inside Iran's border with Turkey, and Ankara is not inclined toward amnesty for these activists.[29] With rare exceptions, the PKK has not been actively involved in urban terrorism inside Turkey, despite Turkey's vulnerability to attacks threatening business and tourism. The potential for a hard core of residual PKK militants to move in this direction cannot be ruled out, although such a campaign would face constraints, including opposition from moderate Kurds living in Turkey's larger cities.

With the containment of the PKK challenge, Turkey now faces the much harder problem of how to resolve its long-standing and troubled relationship with the Kurds by political means. As many as 15 million Kurds live in Turkey, but after years of conflict and the lure of better economic opportunities elsewhere, perhaps no more than 50 percent live in the traditionally Kurdish areas of Southeastern Anatolia. Istanbul, Ankara, and Izmir now have large Kurdish populations, and these populations are highly assimilated—perhaps a quarter of the deputies in the Turkish parliament are of Kurdish origin. But these figures should not be taken to suggest that the pressure for Kurdish cultural and political rights has abated. Although most Kurds, and especially those living in Western Turkey, would not favor the establishment of a separate Kurdish state, the desire for greater cultural and linguistic rights, and perhaps greater autonomy in predominantly Kurdish areas, is strong. The conflict in the Southeast has, if anything, heightened the sense of Kurdish identity inside Turkey, and the Kurdish issue is unlikely to fade even with the decline of the PKK.

Many analysts believe that Öcalan's capture offered an unprecedented opportunity for the development of a new political approach

---

[29]"Turkey and the Kurds: Into a New Era," *IISS Strategic Comments*, Vol. 5, Issue 3, April 1999.

toward the Kurds. To date, Ankara has been unable to seize this opportunity, but the broad possibility of a political settlement remains open, whether through unilateral steps by Ankara or through the cultivation of new, nonviolent Kurdish interlocutors.[30] In the end, any strategy for resolution will involve some concessions from Ankara, and a diminution, however modest, of the sovereignty of the "strong state." The prospects may well depend on the extent of other pressures for modernization and liberalization in Turkey which, if potent enough, could create a more encouraging climate for resolution of the Kurdish problem. In the absence of such a settlement, the issue is likely to remain a central feature of Turkish politics, as well as a domestic security problem and will continue to impose costs on Ankara's international relationships.

ISLAMISM, NATIONALISM, AND TURKISH IDENTITY

Much Western debate about Turkey's internal scene and its foreign policy meaning has focused on the question of Islamism. This was a natural product of the striking rise in the electoral fortunes of the Refah Party in the mid-1990s, first at the local and finally at the national level, leading to the formation of a Refah-led coalition in 1996. In that period, secularism in Turkey seemed to be at bay, and Turkish society was increasingly polarized along religious and secular lines. If there was a prospect for a fundamental reorientation of Turkish external policy, it appeared to arise from the Islamist phenomenon. In retrospect, these concerns proved exaggerated. The Refah experiment in government was short-lived, and Prime Minister Erbakan's forays into foreign policy, including a tour of Islamic capitals (no Arab states were included) and consultations in Libya, were not taken seriously in Turkey or the West. Turkey's 1999 general elections resulted in a poor showing by the Islamists at the national level and a surprisingly strong performance by Turkey's MHP. This result can be ascribed, in part, to the legal measures taken against key Islamist politicians and the closure of Refah. But it also seems that Turkey's nationalists, who also champion the sort of cultural conser-

---

[30]See Philip Robins, "Turkey and the Kurds: Missing Another Opportunity?" in Morton Abramowitz, ed., *Turkey's Transformation and American Policy*, New York: The Century Foundation Press, 2000, pp. 61–93.

vatism favored by the Islamists, have inherited the mantle of opposition to Turkey's traditional politics.

Indeed, Refah's appeal in the mid-1990s also went beyond the issue of religion to include economic populism, anticorruption, and not least, Turkish nationalism. Refah was especially adept at articulating the nationalist message against a background of crises in Bosnia, Chechnya, and Azerbaijan that suggested Western indifference to Turkish and Muslim interests. In the government that emerged after the 1999 elections, with the right-wing MHP as a key partner in Prime Minister Ecevit's coalition, and with MHP's leader, Devlet Bahçeli as Deputy Prime Minister—a strange coalition of the left and the right—the nationalists exerted considerable influence over Turkish policy. MHP is the leading obstacle in most areas to the implementation of Turkey's reform and recovery plans.

Despite MHP's reputation for radicalism and even violence in previous decades, Bahçeli has adopted a relatively moderate line. But some Turkish observers fear that with its growing influence, MHP will sooner or later take a more assertive course. The foreign policy consequences of this could be significant, although the traditional foreign and security policy establishment remains wary of MHP. Under the right conditions, including a new economic shock and the frustration of Turkey's European aspirations, MHP's strongly nationalist line could reinforce tendencies evident elsewhere across the Turkish political spectrum (e.g., Ecevit himself is well known as a nationalist of the left, and many of Turkey's Islamists are highly nationalistic, as noted above).

The areas most sensitive to the policy influence of Turkish nationalism are Cyprus, the nationalist issue par excellence; the Balkans, where Turkey could opt for a more independent and assertive stance; and relations with Russia in the Caucasus and Central Asia. A more nationalist line would have implications for the overall outlook for relations with Greece, especially in the context of a deterioration in relations with the EU. The nationalist impulse would also complicate security cooperation with Washington. MHP has been among the most critical of American use of Incirlik for Operation Northern Watch over Iraq and would be sensitive to policies that cut against Turkey's perceived interests in Iran and Iraq. The imposition of an economic reform package dictated, in large measure, by international financial institutions and Turkey's Western partners has also fanned nationalist opposition.

Turkey's EU candidacy is a significant development, but it clearly does not end the debate over Turkish identity. In some respects, the post-Helsinki environment has actually sharpened the questions of what it means to be a Turk, how Turks view the future of their society, and whether closer integration in Western institutions is compatible with Ankara's more activist approach to international policy. There is a deep reservoir of nationalist sentiment, evident even in mainstream political discourse.[31] Turkey may well continue to evolve along modernist and liberal lines, with a reduced role for the state and a multilateral approach to foreign policy. Under less favorable conditions of economic stringency, regional tensions, or political instability, the departure is more than likely to be in a nationalist direction.

DOMESTIC STABILITY AND TURKEY'S FOREIGN POLICY POTENTIAL

The close connection in the Kemalist outlook between internal stability and foreign policy potential is very evident in the current environment. The connection is neither unusual nor new, but it takes on special meaning in the context of Turkey's expanded external policy horizons and great economic uncertainty. The *tour d'horizon* of regional and functional challenges and opportunities facing Turkey, and discussed in detail in subsequent chapters of this book, raises fundamental questions about Turkey's capacity to respond. Turkey's potential as a "big emerging market," as a positive regional actor, and as an effective partner for Europe and the United States, as well as the pursuit of Ankara's own regional objectives, will depend to a great extent on the energy Turkey can devote to foreign policy and international perceptions of the country's stability and direction. Three issues will be critical in this regard.

---

[31]For an interpretation of Turkish nationalism as a pervasive or "hegemonic" ideology of choice, see Ian Lustick, "Hegemony and the Riddle of Nationalism," in Leonard Binder, ed., *Ethnic Conflict and International Politics in the Middle East*, Gainesville, FL: University Press of Florida, 1999.

First, there are strong pressures for modernization and liberalization at work in Turkey, but ultimately their success will be measured in terms of their corrosive effect on the "strong state," with its ideological and bureaucratic underpinnings. A modern foreign policy, including convergence with European practices, will require sovereignty compromises and a reduction in the pervasive role of the state. In the absence of change in this area, Turkey will be constrained in its foreign and security policy options, and opportunities for expanded economic cooperation may also be limited. Successful economic and political reform can encourage movement in this direction, but the outlook is far from certain.

Second, many Turkish and foreign observers view the current fragmentation of the Turkish political scene, and the rise of movements on the extremes, as inherently unstable and unhealthy for the country's evolution toward a liberal order. A renewal of the center in Turkish political life will almost certainly require modernization and restructuring of the traditional centrist parties, ANAP and DYP, discredited by scandals and political infighting, or the emergence of credible alternatives. Above all, it is likely to require the emergence of a new generation of capable leaders to replace an aging political class, possibly drawn from other-than-traditional party circles. In this context, some observers see the emergence of Mehmet Ali Bayar as head of the Democratic Turkey Party—an amalgam of centrist, populist visions in the Demirel vein—as a promising new force in Turkey's politics. In the absence of competent "new faces" at the center, Turkey's political scene may remain uncertain, with the potential for radical departures in foreign and security policy, and very negative consequences for foreign investment and Ankara's regional and transatlantic relationships.

Third, Turkey's considerable economic success of the past decades, followed by a virtual collapse in 2000–2002, raises important questions of equity and social cohesion. Turkey is certainly not alone in facing this developmental dilemma. But as a "pivot" state, the stakes are relatively high for Turkey and its international partners. With other states around the Southern Mediterranean, Turkey suffers from a growing problem of income disparity. Persistent high rates of inflation, and the devaluation of 2001, have had a disproportionate effect on the country's large but insecure middle class. The by-products of Turkey's economic growth are all too visible to the vast majority of Turks who do not participate in the country's financial markets and have to live with minimal social services. It is precisely these conditions that have fueled religious politics, especially at the local level, and not just among Turkey's poor but among many middle class voters. The problem of corruption is part of this equation to the extent that it plays a large role in public perceptions about the Turkish economy and the quality of Turkish governance.

In important respects, Turkey is at a turning point in its evolution as a state and society. The direction Turkey takes on key questions such as the role of the state, civil-military relations, economic reform, secular versus religious politics, and a nationalist versus internationalist orientation will have a strong effect on the country's power and potential in the coming years. In an era in which the ability to hold down Soviet divisions in Thrace and the Caucasus is no longer a valid measure of international weight, or a basis for alliance relationships, Turkey's internal evolution matters a great deal and will influence the character and extent of European and American engagement with Ankara. Turkey has already emerged as an important regional and "trans-regional" actor. Its capacity to sustain and expand this role will depend on Turkey's internal development as much as changes in the external environment. But to return to Atatürk's observation offered at the beginning of this chapter, the two spheres are interdependent—increasingly so as public opinion has become more aware of international issues and as the number and diversity of actors involved in Turkey's foreign policy debate has grown.

Chapter Three

TURKEY AND EUROPE

Diplomatically, Turkey has been part of the European state system since the 19th century when the Ottoman empire was included in the Concert of Europe. At the Paris Peace Conference in 1856, Europe's great powers decided that the territorial integrity of the Ottoman empire was essential for European stability. Indeed, for much of the last half of the 19th century, European diplomacy was dominated by the "Eastern Question"—that is, how to manage the decline of the Ottoman empire, which by the mid-1800s had become, in Czar Nicholas I's famous phrase, "the sick man of Europe."

Yet although the Ottoman empire was part of the European state system, it was never regarded as an equal member of it. At its root, the Concert of Europe was an association of Christian, European, and "civilized" states governed by certain values and norms. Most European statesmen felt that the Ottoman empire did not share these values and norms. As Iver Neumann has noted, "Although 'the Turk' was part of the system of interstate relations, the topic of culture denied it equal status within the community of Europe."[1]

Turkey's relations with Europe changed with the demise of the Ottoman empire. As the Ottoman empire lost its military superiority and fell behind European states in technological development, the Ottoman elite began to look to Europe as a model and began to import European ideas, lifestyles, and ways of thinking. In the words of one Turkish scholar, Europe became "a mirror through which the Ot-

---

[1]See Iver B. Neumann, *Uses of the Other: The "East" in European Identity Formation,* Minneapolis: University of Minnesota Press, 1999, p. 59.

toman elite perceived its own weaknesses, differences and traits."[2] The "Europeanization" process thus became critical in defining Turkey's development and evolution.

This process of modernization and Europeanization was accelerated in the early 20th century, especially after the formation of the Turkish Republic by Kemal Atatürk in 1923. Atatürk sought to modernize Turkey by eliminating almost all aspects of the Ottoman system, which he regarded as the main reason for Turkey's cultural and technological backwardness. Beginning with the abolition of the Sultanate in 1923, and the Caliphate a year later, Atatürk introduced a series of reforms designed to transform Turkey into a modern, secular European state.[3]

This process of gradual Europeanization was given new impetus after World War II by Turkey's entry into NATO. Turkey was regarded as an important bulwark against the expansion of Soviet power and a critical link in the Western defense system. In 1963, Turkey concluded an Association Agreement (the Ankara Agreement) with the European Community (EC), which foresaw the possibility of eventual membership (Article 28) once the conditions for membership had been met.

At the same time, there has always been—and continues to be—a sense among many Europeans that Turkey is not really "European." For centuries, "the Turk" was the significant "Other" against which Europe defined its identity.[4] This perception of Turks as "other" in Europe is deeply embedded in Europeans' collective memory and colors European views of Turkey today. Because of its different cultural and religious traditions, Turkey is seen as not quite really "European." As one former European Union official has put it, "Turkey has never been fully considered a European country, but

---

[2]Meltem Müftüler-Bac, "Through the Looking Glass: Turkey in Europe," *Turkish Studies*, Vol. 1, No. 1, Spring 2000, p. 28.

[3]For a detailed discussion, see Bernard Lewis, *The Emergence of Modern Turkey*, London: Oxford University Press, 1961, especially pp. 234–293.

[4]For a detailed exploration of this theme, see Neumann, *Uses of the Other*, Chapter Two.

neither is it considered fully Asian. It is at the crossroads between two continents, two cultures and two destinies."[5]

This ambiguity about Turkey's place in Europe—its "European-ness"—has become more acute since the end of the Cold War. As long as the Soviet Union was perceived as a major threat, strategic considerations tended to dominate Turkey's relationship to Europe. Although many Europeans had doubts about whether Turkey could ever become a member of the European Community—as the EU was then called—these doubts took a backseat to the overriding strategic need to bind Turkey close to the West.

The end of the Cold War, however, has raised new doubts about Turkey's place in Europe and created new difficulties in Ankara's relations with Europe. With the demise of the Soviet Union, military strategic considerations have become less important in Europe's approach to Turkey, whereas economic, political, and cultural factors have increased in importance. Today Europe is not primarily concerned about deterring a Soviet (or Russian) threat but about creating a cohesive political and economic union and forging an effective common European foreign and defense policy. This shift in priorities has highlighted Turkey's "distinctiveness" and raised new questions about where Turkey fits into the "new Europe."

TURKEY AND EC ENLARGEMENT

Turkey's problems with Europe—and the European Union in particular—have been accentuated by the EC's (and later the EU's) changing approach to enlargement.[6] The Association Agreements which the EC signed with Greece in 1961 and Turkey in 1963 provided important trade benefits and were seen as possible stepping stones to full membership if and when the two countries fulfilled the concrete conditions for membership.

---

[5]See Eberhard Rhein, "Europe and the Greater Middle East," in Robert D. Blackwill and Michael Stürmer, eds., *Allies Divided: Transatlantic Policies for the Greater Middle East*, Cambridge, MA: MIT Press, 1997, p. 47.

[6]For good discussion, see Sevilay Elgün Kahraman, "Rethinking Turkey-European Union Relations in Light of Enlargement," *Turkish Studies*, Vol. 1, No. 1, Spring 2000, pp. 1–20.

At the time of their conclusion, the Association Agreements were essentially limited to trade and financial matters. However, over the next several decades the EC's goals and competence expanded significantly. First, Turkey's economic development, which until the early 1980s relied heavily on an industrialization strategy based on import substitution, was in conflict with its commitment to economic integration with the EC through a customs union. This contradiction inhibited the development of Turkey's relations with the EC.

In addition, Turkey's protectionist strategy separated Turkey from the pattern of economic development in the rest of Europe, especially Southern Europe (Spain, Portugal, and Greece). Whereas the countries of Southern Europe experienced a rapid growth of imports and revenue from tourism, Turkey experienced no such boom.[7] Thus, although Turkey's relations with the EC in the 1970s were characterized by increasing tensions over the lack of progress toward the creation of a customs union, Greece, Spain, and Portugal succeeded in intensifying their efforts to integrate themselves into the European core.

The growing integration of the three Southern countries into the European core also led to an increasing "Europeanization" of their foreign policies—a process that gained greater impetus after their entry into the EC. This Europeanization process, however, did not occur in Turkey. Ankara continued to orient its policy more toward Washington than Brussels. Thus, although the economic and political aspects of Europeanness complemented one another in the case of Southern Europe, this did not happen in Turkey's case. Turkey remained outside the process of Europeanization that increasingly characterized the economic and political evolution of the other South European countries.

Moreover, the Southern enlargement that resulted in the entry of Greece (1981), Spain (1986), and Portugal (1986) into the EC reflected an important shift in the EC's approach to enlargement. In opening up its ranks to the three South European countries, the EC gave priority to political considerations—particularly the desire to

---

[7] Sevilay Elgün Kahraman, "Rethinking Turkey-European Union Relations in Light of Enlargement," p. 3.

stabilize democracy in these countries—over economic concerns. It thus introduced additional criteria for membership for future members such as adherence to democratic principles, respect for human rights, and the rule of law. In effect, as Sevilay Elgün Kahraman has noted, the Southern enlargement resulted in "a reformulation of the external identity of the community."[8] Rather than being primarily an economic club of Northern industrialized countries, the EC came to represent shared norms, values, and codes of behavior among its members.

This shift in emphasis in EC policy went largely unnoticed in Turkey. Instead, Turkey continued to emphasize the *economic aspects* of membership, especially after the introduction of free-market reforms by Prime Minister Turgut Özal in the early 1980s. In 1987, Turkey applied for membership, despite being cautioned against doing so by the EC. However, unlike Greek Prime Minister Constantine Kara-manlis, Özal did not carefully prepare the groundwork for Turkey's application, either at home or in Brussels.

Moreover, the timing was bad. The EC had only just begun the diffi-cult process of digesting the Southern enlargement (Greece, Spain, and Portugal). In addition, the EC was on the verge of trying to de-velop a common internal market by 1992 and streamlining its deci-sionmaking processes. Thus, the EC was not ready to begin another round of enlargement, especially one involving membership of a large and less economically developed country like Turkey, which would entail significant financial burdens.

In December 1989 the EC rejected Turkey's membership, citing a va-riety of economic, social, and political reasons.[9] Instead, the EC pro-posed an intensification of relations based on the existing Associa-

---

[8]Ibid., p. 5.

[9]See *Commission Opinion on Turkey's Request for Accession to the Community* [SEC (89) 2290 fin./2], Brussels, December 20, 1989. For a detailed discussion of Turkish and EC considerations regarding Turkey's application, see Heinz Kramer, *Die* Eu-ropäische *Gemeinschaft und die Turkei: Entwicklung, Probleme und Perspektiven einer schwierigen Partnerschaft*, Baden-Baden: Nomos Verlagsgesellschaft, 1988, pp. 84–111 and 120–150. See also Heinz Kramer, "Turkey and the European Union: A Multi-Dimensional Relationship," in Vojtech Mastny and R. Craig Nation, eds., *Turkey Between East and West: New Challenges for a Rising Regional Power*, Boulder, CO: Westview Press, 1996, pp. 203–232.

tion Agreement. In line with this, the EC Commission presented a comprehensive package of economic, trade and political measures—the so-called Matutes Package—designed to improve EC-Turkish relations. These measures were designed to relaunch EC-Turkish relations, which had largely languished after the military coup in Turkey in 1980.

THE IMPACT OF EASTERN ENLARGEMENT

Turkey's bid for EC membership was further complicated by the collapse of communism in Eastern Europe in 1989–1990. This created a new dilemma for the EC: how to facilitate the "return to Europe" of the countries of Eastern Europe which had just emerged from 45 years of communist rule. Suddenly, Turkey found itself thrust to the back of the enlargement queue by the emergence of a large new group of candidates for membership—countries that only a few years earlier had been on the other side of the East-West divide. Although many of the East European countries were less advanced economically than Turkey, they were considered to be politically and culturally a part of Europe. Thus, the EC's attempt to integrate the East European countries added a new "cultural dimension" to the EC's policy that previously had been absent.

The Copenhagen summit in June 1993 was an important watershed in the evolution of the EU's approach to enlargement.[10] First, it acknowledged that membership of the East European and Baltic countries—but not Turkey—was a major objective of EU policy. Second, it set out specific economic *and political* criteria for membership. Among the latter were the requirement that a candidate achieve a stable democracy, maintain the rule of law, respect human rights, and protect the rights of minorities.

The Copenhagen criteria significantly complicated Turkey's quest for EU membership. In effect, they raised the bar for membership by adding a whole new group of political criteria that had previously not been explicit criteria for membership. In so doing, they accentuated the difference between Turkey and the other aspirants and highlighted Turkey's "distinctiveness." Whereas Turkey's economic

---

[10]The European Community changed its name to the European Union in 1993.

qualifications were better than those of many of the Central and East European states, Turkey fell short on many of the political criteria, especially those related to human rights.

The EU's Luxembourg summit (December 1997) essentially ratified this process of differentiation. While underlining the gradual and all-inclusive nature of the enlargement process, it set up a two-tier accession process with the 11 East European and Baltic countries plus Cyprus. Turkey, however, was not accepted as a candidate country. Instead, a "pre-accession" strategy was set out for Turkey designed to help it to enhance its candidacy for membership.

In other words, Turkey was included in the enlargement process but *not in the pre-accession strategy* along with the other Central and East European countries (plus Cyprus). Instead, it was offered a special strategy to help it prepare for accession, but it was not given official candidate status nor offered a timetable for accession negotiations. The EU also linked Turkey's eligibility as a candidate to progress on issues not contained in the Copenhagen criteria such as a resolution of Turkey's differences with Greece over the Aegean and a Cyprus settlement.

The EU's failure to include Turkey on the list of candidate countries provoked a wave of outrage in Turkey and prompted Turkey to freeze its political dialogue with the EU. Many Turks believed—and continue to believe—that the EU's rejection of Turkey's candidacy was unfair and reflected an inherent bias against Turkey on cultural and religious grounds because it is a Muslim country. Although such concerns are not entirely absent from European thinking, this view overlooks the EU's evolution in the last decade and the hardening of accession criteria, particularly the growing importance of political criteria for membership. It is this evolution and greater emphasis on political criteria more than anything else that has complicated Turkey's efforts to achieve EU membership.

THE HELSINKI SUMMIT

The Luxembourg summit represented a nadir in Turkey's relations with the EU. In the aftermath of the summit, Turkish-EU relations sharply deteriorated and Turkey downgraded its relations with Brussels. Indeed, in the wake of the summit, it was common to hear voices in Ankara and Istanbul saying that Turkey's relations with Europe were not all that important and that Turkey needed to diversify its ties. However, at the Helsinki summit (December 1999), the EU reversed its previous stand and accepted Turkey as a candidate member "on the basis of the same criteria as applied to the other candidate states."[11] Turkey was also offered a pre-accession strategy designed to stimulate and support its reforms.[12]

The shift in the EU's position was due to several factors: (1) a desire on the part of the EU to halt the deterioration of Turkish-EU relations after Luxembourg, (2) a more accommodating position by the new SPD/Green coalition in Germany, (3) a change in Greek policy toward Turkey, and (4) pressure from the United States. However, the EU made clear that it was not prepared to open accession negotiations with Turkey until Ankara had fulfilled the Copenhagen criteria. It also linked Turkey's eventual membership to a resolution of its dispute with Greece over the Aegean and a settlement of the Cyprus issue.[13]

At its summit in Nice (December 4–6, 2000), the EU Council approved an Accession Partnership Document for Turkey. The Accession Partnership (AP) is the centerpiece of the pre-accession strategy for Turkey. It identifies short- and medium-term priorities, intermediate objectives, and aspects on which Turkey's accession preparations must concentrate for Turkey to qualify for membership. It

---

[11]For a detailed discussion of developments leading up to the change in the EU position, see Ziya Onis, "Luxembourg, Helsinki and Beyond: Towards an Interpretation of Recent Turkey-EU Relations," *Government and Opposition*, Vol. 35, No. 4, Autumn 2000, pp. 403–483; and William Park, "Turkey's European Union Candidacy: From Luxembourg to Helsinki—to Ankara?" *Mediterranean Politics*, Vol. 5, No. 3, Autumn 2000, pp. 31–53.

[12]*Presidency Conclusions,* Paragraph 12, Helsinki European Council, December 10–11, 1999, available at http://europa.eu.int/council/off/conclu/dec99/dec99_en.htm.

[13]Some Turkish analysts have suggested that the reason why the EU accepted Turkey's candidacy at the Helsinki summit was related to the geopolitical and security risks that might arise if Turkey were excluded. (See, in particular, Meltem Müftüler-Bac, "Turkey's Role in the EU's Security and Foreign Policies," *Security Dialogue*, Vol. 31, No. 4, December 2000, pp. 489–502.) However, there is little evidence to support this view. On the contrary, the Helsinki decision appears to have been based on a combination of economic, political, and social considerations. Security concerns played practically no role. For good discussion, see Gareth Jenkins, "Turkey and EU Security: Camouflage or Criterion for Candidacy?" *Security Dialogue*, Vol. 32, No. 2, June 2001, pp. 269–272.

also sets up monitoring mechanisms to evaluate progress toward achieving the objectives and priorities set out in the Accession Partnership Document. In effect, the Accession Partnership Document establishes a "roadmap" for Turkish accession to the EU. Whether—or when—Turkey joins the EU depends critically on Turkey's ability to realize the objectives set out in the Accession Partnership Document.

The National Program of Action presented by the Turkish government in March 2001 was supposed to spell out how Ankara intended to meet the objectives laid out in the Accession Partnership Document. The program was an impressive declaration of intentions, but it was vague and evasive on precisely those issues—the treatment of minorities, the role of the military, Cyprus, and relations with Greece—regarded by the EU as the most critical areas where changes are needed.[14]

Since then, Turkey has taken a number of steps to address EU concerns. In October 2001, the Turkish parliament passed a series of reforms that significantly eased restrictions on human rights.[15] These reforms range from reducing police powers of detention to easing the investigation of legislators suspected of corruption and other crimes. The reforms also paved the way for lifting the ban on Kurdish language broadcasts and increased civilian representation on the National Security Council. In addition, in February 2002, the parliament passed a mini-reform package containing reforms in the Turkish Penal Code and antiterrorism law that relaxed constraints on freedom of expression that had been used to jail journalists and intellectuals who published views considered to undermine the State.

These reforms have been welcomed by the EU. However, the Progress Report released by the EU in October 2002 makes clear that Turkey still has a way to go to fulfill the Copenhagen criteria,

---

[14]See "Program falls short of expectations," *Turkish Daily News*, March 20, 2001. Also "Harte Kritik an Ankaras EU-Reformprogramm," *Neue Zürcher Zeitung*, March 21, 2001; and "EU tells Turkey to do more," *Turkish Daily News*, March 27, 2001.

[15]See Leyla Boulton, "Turkey approves reforms to ease curbs on human rights," *Financial Times*, September 24, 2001. Also Rainer Hermann, "Reform in kleinen Schritten," *Frankfurter Allgemeine Zeitung*, October 30, 2001.

especially in the political field.[16] In addition, Ankara must overcome a number of other obstacles—including resolving its differences with Greece over the Aegean and Cyprus—before membership in the EU can seriously be considered.

THE ECONOMIC DIMENSION

The most important obstacles to Turkey's EU membership are political, but there are important economic obstacles to membership as well. One is simply the huge size of the country. With a population of nearly 67.8 million, Turkey is the second largest country in Europe behind Germany. If its population continues to grow at the current rate, Turkey could have the largest population in Europe by the middle of the 21st century. Integrating a country the size of Turkey—especially one that is also characterized by great regional disparities—will pose an enormous challenge to the EU.

The reforms introduced by Prime Minister Özal in the early 1980s essentially abandoned the import substitution strategy that had previously been the cornerstone of Turkish economic policy. Since then, Turkey's economy has become more open and the private sector has expanded significantly. Currently, more than half of Turkey's foreign trade is with the EU. Moreover, the structure of Turkey's trade with the EU has significantly changed. In the 1970s, Turkey was an exporter of agricultural produce and raw materials. Today, the bulk of Turkey's exports are manufactured goods.

Despite these changes, the Turkish economy is still characterized by a number of structural weaknesses that inhibit Turkey's integration into the EU:

- **Low Per Capita Income.** Income levels in Turkey are significantly below those in Western Europe. Per capita Gross Domestic Product (GDP) is about $3,000. This is well below that of the poorest countries in Europe. Greece and Portugal, for instance, have per capita incomes of $11,770 and $10,600, respectively.

---

[16]See *Regular Report on Turkey's Progress Toward Accession*, Brussels, Commission of the European Communities, SEC (2002), 1412, October 10, 2002..

- **A Large Agricultural Work Force.** Nearly 40 percent of Turkey's population is engaged in agriculture (compared with 2 percent in Britain). This rural work force amounts to about 15 million people—a number larger than the total population of several current EU members and five times larger than the population of some prospective new members such as Slovenia and Estonia. Shifting this population out of agriculture into modern forms of economic activity will require an enormous and prolonged structural adjustment.

- **Large Regional Disparities.** Although areas of Istanbul and Ankara enjoy standards of living and levels of prosperity close to those of Western Europe, Central and Eastern Anatolia are largely agricultural and have a much lower standard of living. With the winding down of the PKK insurgency, however, the Turkish government has begun to encourage greater private investment in Southeast Anatolia.

- **High Inflation.** Turkey's inflation over the last decade has been nearly 80 percent per annum—much higher than the average in the EU. This has had a negative effect on the financial sector and is one reason for Turkey's low foreign investment. Lower inflation would not only increase foreign investment but could also help to boost Turkey's growth rates.

- **Low Foreign Investment.** Turkey has a very low rate of foreign investment. In the period 1993–1997, foreign investment averaged about 0.5 percent of Gross National Product (GNP). By contrast, during the same period Spain, Greece, and Portugal managed to attract foreign investment of between 1 and 2 percent of GNP. The low rate of foreign investment is due largely to macro-economic instability and regulatory deficiencies. But other factors such as cumbersome bureaucratic practices and the persistent high inflation rate have also played a role in inhibiting foreign investment.

- **A High Public Sector Deficit.** Turkey's public sector deficit is extremely high—35–45 percent of the GNP. This is one factor responsible for Turkey's high inflation. This high inflation rate is compounded by structural weaknesses in the financial sector that make it difficult for Turkish banks to compete with their counterparts in the EU.

- **Slow Pace of Privatization.** Turkey has undertaken a major effort at privatizing state-owned enterprises since the early 1980s. However, the pace of privatization has been sluggish and hampered by the fact that not all of the necessary legislation has been passed. In addition, a lack of transparency in decisionmaking has resulted in large-scale fraud, bribery, and corruption.

In short, Turkey faces major problems of structural adjustment before it is ready for membership in the EU. This is true, above all, in the agricultural sector, which, as noted above, still accounts for 40 percent of the workforce. The creation of the customs union, which came into force at the end of 1995, should help reduce many of these structured obstacles. The customs union is designed to abolish tariffs on imports. It initially resulted in a worsening of Turkey's balance of trade with the EU, but over the longer term it will create a more liberalized economic climate and should help Turkey to integrate into the global market.

At the same time, the economic crisis of 2000–2002 has underscored the need to proceed more rapidly with a program of structural reforms. Reforms are required in three areas in particular: agriculture, energy, and privatization. Reform in these areas will help to restore investor confidence—badly shaken by the economic crisis—and help reduce corruption by increasing transparency. It will also enhance Turkey's prospects for EU membership over the long term.

DEMOCRACY AND HUMAN RIGHTS

The main obstacles to Turkey's EU membership, however, are in the field of democratization and human rights. Recent EU evaluations of Turkey's prospects for membership have consistently pointed to shortcomings in Turkey's human rights record.[17] Although Turkey has taken some steps to address these shortcomings, progress has been slow and insufficient. Moreover, many of the changes in legislation have not been implemented.

One of the most sensitive issues is the use of torture in Turkey. At the Luxembourg summit in December 1997, Luxembourg Prime Minister

---

[17]Ibid.

Jean-Claude Juncker called attention to this issue by bluntly stating that torturers could not sit at the EU table. Juncker's remarks caused uproar in Turkey—in part because of the undiplomatic manner in which they were expressed. They nonetheless highlight the need for Turkey to address the torture issue more adequately if it is to achieve its goal of EU membership.

Two points, however, need emphasis in this connection. First, although there continue to be deficiencies in Turkey's record regarding the use of torture, these lapses are largely a result of *poor implementation* of the laws and regulations on the books, not the result of a state policy that condones torture. Second, in the last several years Turkey has undertaken a number of efforts to tighten the laws against torture. However, the fact remains that incidents of torture continue to occur.[18] Thus, Turkey needs to take further steps to ensure that the laws and regulations against the use of torture are fully implemented if it expects to gain membership in the EU.

The EU has also expressed concern about restrictions on freedom of expression, especially those contained in Articles 159 and 312 of the Turkish Penal Code and paragraphs 7 and 8 of the Turkish Anti-Terrorism Law. In February 2002, after intense debate, the Turkish parliament passed legislation amending these laws. However, the EU has claimed that these changes do not go far enough to meet the Copenhagen criteria.[19]

In an effort to further enhance Turkey's qualifications for EU membership, in August 2002 the Turkish parliament passed a major reform package that *inter alia* abolished the death penalty except in times of war. The legislation removed an important obstacle to EU membership. However, it is not likely to be enough to persuade the

---

[18]The European Committee Against Torture published three reports in 1992, 1996, and 1999 in which it noted that torture was systematically practiced in Turkey. The most recent report, published in February 1999, noted that the situation was improving but stated that torture still occurs in Turkey and that, despite the improved rules and regulations forbidding torture, the implementation of the rules and regulations was not satisfactory, especially in police stations and prisons. See *Toward Calmer Waters: A Report on Relations Between Turkey and the European Union*, The Hague: Advisory Council on International Affairs, July 1999, p. 28.

[19]Selcuk Gültasli, "EU: 312 and 159 amendments not sufficient," *Turkish Daily News*, February 8, 2002.

EU to open accession negotiations at the EU Copenhagen summit in December 2002.

THE KURDISH ISSUE

A related problem is posed by Turkey's approach to minority rights, especially for the Kurdish population. Kurds comprise between 8 and 12 million people out of Turkey's total population of nearly 67.8 million. Less than half the Kurdish population is in Southeast Anatolia; the majority is in the major Turkish cities, particularly Istanbul and Ankara. Another eight million or so Kurds live in the Kurdish regions of Iraq, Iran and Syria.

Turkey does not recognize a Kurdish minority and views Kurds simply as citizens of Turkey—in short, Turks. This is a legacy of the strong assimilationist policies pursued by Atatürk at the time of the founding of the Turkish Republic in 1923. The Ottoman empire was organized along religious community lines. Thus, the Kurds—unlike the Greeks, Jews, and Christian Armenians—were not considered a minority but were full members of the Muslim majority. Atatürk was determined to create a new Turkish nation-state on the basis of a specifically "Turkish" national identity. As a result, all existing Muslim minorities, including the Kurds, were "granted a kind of Turkishness."[20] The only minorities that were recognized were those recognized in the Lausanne Treaty (1923)—Jews, Armenians, and Greeks.

Atatürk's concept of Turkish identity was highly inclusive. Every person living within the borders of the Turkish Republic and accepting its basic principles could become a Turkish citizen. But becoming a Turk required the suppression of an individual's ethnic identity. In short, Atatürk's concept was extremely generous in that it allowed anyone to become a Turkish citizen. But as Svante Cornell has pointed out, it did not provide a solution to those who were not pre-

---

[20]Dogu Ergil, "Identity Crisis and Political Instability in Turkey," *The Journal of International Affairs*, Vol. 54, No. 1, Fall 2000, p. 51. However, by forcing all Muslims into a Turkish identity, the new regime also closely associated Turkish identity with Islam, which was contrary to its secularization policy.

pared to abandon their previous ethnic identity.[21] This was the case for a large portion of the Kurdish population.

The Turkish approach to the Kurdish issue is also animated by deep-seated historical memories regarding threats to the integrity of the Ottoman empire at the end of the 19th century and reflects a long-standing belief, sparked in particular by the Sèvres Treaty (1920), that the West favors the dismemberment of the Turkish state and the creation of an independent Kurdish state.[22] As a result, Turkey has strongly rejected demands by the Kurds for greater regional autonomy and cultural rights, such as the right to receive public education in their own language, fearing that this could spark separatist pressures and threats to the integrity of the Turkish state.

The winding down of the campaign against the PKK has removed an important point of tension in Turkey's relations with the EU. At the same time, it has focused greater attention on Turkey's internal policies and made it harder to justify some of the restrictions on personal expression and human rights. These restrictions were justified in the past on the grounds that Turkey faced a serious terrorist and separatist threat. But now that the threat from the PKK has significantly diminished, the case for keeping these restrictions in place is much less compelling.

The diminution of the armed violence by the PKK provides an opportunity for the Turkish government to press forward with reforms in the field of democratization and human rights. The government has recently undertaken measures to encourage investment in the Kurdish-populated areas of Southeastern Anatolia, one of the poorest regions in Turkey. In August 2002, in an effort to encourage the EU to open accession negotiations, the Turkish parliament also passed legislation legalizing broadcasting and private tutoring in

---

[21]See Svante Cornell, "The Kurdish Question in Turkish Politics," *Orbis*, Vol. 45, No. 1, Winter 2000, p. 34.

[22]Under the Sèvres Treaty, the Western powers decided to dismantle the Ottoman empire and agreed to the establishment of an independent Kurdistan. Although the treaty was never implemented, the memory of the treaty has had a strong effect on Turkey's national consciousness and psyche.

Kurdish. However, teaching Kurdish in public schools is still forbidden.[23]

These moves remove important obstacles to EU membership. But they are unlikely to be enough to persuade the EU to open accession negotiations at the EU summit in Copenhagen. The best Turkey can probably hope for at the summit is a vague reference to the EU's willingness to open accession negotiations when Turkey has fulfilled the Copenhagen criteria.

THE ISLAMIC FACTOR

Turkey's application for EU membership also raises important cultural and "civilizational" issues. Although the EU insists that Turkey's application for membership will be judged solely on the Copenhagen criteria, beneath the surface many Europeans question the degree to which Turkey's Islamic religious and cultural traditions are compatible with "European" values. As former Dutch Foreign Minister Hans van Mierlo stated in early 1997: "There is a problem of a large Muslim state. Do we want that in Europe? It is an unspoken question."[24]

Over the past several decades, there has been a visible increase in the role of Islam in Turkish social and political life. This has been reflected, in particular, in the strong support for the Islamic Refah (Welfare) Party and its successor, the Fazilet (Virtue) Party. It has become fashionable lately among Western observers to talk about a process of "re-Islamization" in Turkey. However, such a view is misleading. Islam was never really eliminated in Turkey. It was simply removed from state institutions. It continued, however, to exert a strong influence in the countryside. The result was a sharp division between the Kemalist secular culture of the military-bureaucratic elite centered in Ankara and other major cities in Western Turkey and the traditional Islamic culture that prevailed in the villages and towns of Eastern Anatolia.

---

[23]Karl Vick, "Turkey Passes Rights Reforms in Bid for EU," *The Washington Post*, August 4, 2002.

[24]Quoted in Stephen Kinzer, "Turkey Finds European Door Slow to Open," *New York Times*, February 23, 1997.

Beginning in the 1950s and 1960s, Islam gradually began to make a comeback in Turkey.[25] This was in part a product of the democratization of Turkish political life, which gave new space to all kinds of political groups, including the Islamists. Ironically, in the 1980s it was also promoted by the military, who saw Islam as a bulwark against the infiltration of Marxist and leftist ideas and made religious instruction in schools compulsory. Both General Kenan Evren, who became president after the return to civilian rule in 1983, and Prime Minister Turgut Özal repeatedly stressed the importance of religious values in Turkish nationalism. Özal even performed the hajj—the holy pilgrimage to Mecca—during a trip to Saudi Arabia.

Broader economic and social changes, particularly increasing urbanization, also contributed to strengthening the role of Islam in Turkish political and social life. As more and more Turkish peasants flocked to the cities, they brought with them their rural values, including their strong Islamic traditions. Uprooted from familiar surroundings and often forced to live in shanty towns (*gecekondu*) on the outskirts of major cities, this disaffected and impoverished stratum of Turkish society has been one of the strongest sources of political support for the religious parties. In effect, Islam has become "the oppositional identity for the excluded sectors of Turkish society."[26]

The Islamists have also benefited from the failure of the mainstream political parties to address Turkey's growing social and economic problems. This has allowed the Islamists to portray themselves as the party of clean, efficient government, especially at the local level.[27] Indeed, many of those who voted for Refah in the 1995 national elections—in which Refah won more votes (21 percent) than any other political party—did so out of disillusionment with the

---

[25]For a detailed discussion, see Jeremy Salt, "Nationalism and the Rise of Muslim Sentiment," *Middle Eastern Studies*, Vol. 31, No. 1, January 1995, pp. 13–27, and Sencer Ayata, "Patronage Party and the State: The Politicization of Islam in Turkey," *The Middle East Journal*, Vol. 50, No. 1, Winter 1996, pp. 40–56.

[26]M. Hakan Yavuz, "Cleansing Islam from the Public Sphere," *The Journal of International Affairs*, Vol. 54, No. 1, Fall 2000, p. 22.

[27]Refah's record for clean, efficient government at the local level was a major factor in its strong showing in the December 1995 elections. However, its policies also had a less-publicized dark side, including mounting debt in many municipalities. See Ugur Akinci, "The Welfare Party's Municipal Track Record: Evaluating Islamist Activism in Turkey," *The Middle East Journal*, Vol. 53, No. 1, Winter 1999, pp. 75–94.

mainstream parties and as a form of protest rather than out of religious conviction.

At the same time, the Islamist movement has undergone an important evolution in recent years.[28] All major Islamic groups have become more "pro-European." Once firm opponents of Turkey's membership in the EU, the Islamists today are one of the strongest supporters of Turkish membership, which they see as an important guarantee of their religious and political rights. In addition, the crackdown on Islamic influences since the ouster of the Erbakan government in mid 1997—the so-called "February 28 Process"—has prompted a rethinking of attitudes toward modernity and democracy within the Islamic movement and the now banned Virtue (Fazilet) Party. In the last few years, a group of younger, pragmatic "modern Islamists" has increasingly challenged the antimodern, dogmatic policies of former Prime Minister Necmettin Erbakan and tried to steer the Islamist movement in a more democratic direction.[29]

The decision by the Constitutional Court in June 2001 to ban the Virtue Party has accelerated the transformation of the Islamic movement and resulted in the emergence of a more democratic and forward-looking Islamist Party—the Justice and Development Party (AKP)—one that could be easier to integrate into Turkish political life.[30] Polls show that the AKP is likely to get more than 20 percent of the vote in the November 2002 elections. Thus, it is quite possible that the AKP could be part of a government that emerges from the elections. Whether the Turkish military would be willing to accept such an outcome or would initiate a campaign to undermine an AKP-led government, as it did in the case of the Erbakan government, remains to be seen.

---

[28]However, the conservative wing of the Islamist movement, led by Erbakan, has been less favorably disposed toward EU membership since the decision by the Turkish Constitutional Court in June 2001 to ban the Virtue Party.

[29]For a detailed discussion, see Chapter Two.

[30]On the prospects for a "reconciliation" between Islam and democracy, see Metin Heper, "Islam and Democracy in Turkey: Toward a Reconciliation?" *The Middle East Journal*, Vol. 51, No. 1, Winter 1997, pp. 32–45. Also, see Resat Kasaba, "Cohabitation? Islamist and Secular Groups in Modern Turkey," in Robert W. Heffner, ed., *Democratic Civility*, New Brunswick, NJ: Transition, 1998, pp. 265–284. For a useful survey of the new pro-Islamic parties in Turkey, see Günter Seufert, *Neue pro-Islamische Parteien in der Turkei*, SWP-Studie, Berlin: Stiftung Wissenschaft und Politik, March 2002.

CIVILIAN CONTROL OF THE MILITARY

Turkey's aspirations for EU membership will also require an important change in the role of the military in Turkish political life. In the postwar period, the Turkish military has acted as custodian of the Kemalist legacy. The military sees its mission as not only to defend the territorial integrity of the Turkish state against external threats but also to protect it against internal challenges.

Ironically, as the Islamists have sought to modernize and move further away from fundamentalism, the military has become more doctrinaire and dogmatic in its interpretation of Kemalism. Although outwardly strong supporters of Turkey's membership in the EU, many officers fear that the reforms required for EU membership will weaken the ability of the Turkish state to manage its security problems—a view shared by the right-wing MHP.[31] Initially, the military opposed any significant relaxation of restrictions on freedom of expression in the Constitution and Penal Code, including easing the ban on broadcasts in Kurdish, arguing that the lifting of these restrictions would pose a threat to the integrity of the Turkish state.[32] However, they have recently taken a more flexible approach to broadcasting in Kurdish and abolition of the death penalty.

The military's special role is codified through its dominance of the NSC. Legally, the NSC has only an advisory function. In reality, however, NSC pronouncements are tantamount to official edicts—as former Prime Minister Erbakan was forced to recognize when he sought to treat the NSC's "recommendation" to clamp down on the rising influence of Islamist forces in Turkey as only "advice." His failure to take forceful action against the Islamists lead to his eventual ouster in a "silent coup" in June 1997.

In response to EU pressure, the Turkish parliament passed legislation in October 2001 calling for an increase in the number of civilians on the NSC. However, this is largely a cosmetic measure. The

---

[31] See the article by the Deputy Chairman of the Nationalist Action Party Sevkot Bülent Yahnici, "EU road might be a trap for ethnic disintegration," *Turkish Daily News*, January 3, 2001. See also Elif Ünal, "Nationalists fear reforms will split Turkey," ibid., January 20, 2001.

[32] See Leyla Boulton, "Turkish Military Resists EU Demands on Kurds," *Financial Times*, December 12, 2000.

influence of the military on many security issues remains strong. Membership in the EU will require a reduction of the military's influence in Turkish politics. Whether the military will be willing to accept such a lower profile remains to be seen.

GREECE AND CYPRUS

Turkey's differences with Greece over the Aegean and Cyprus—discussed in detail in Chapter Five—also are an important factor affecting Ankara's overall relationship with the EU. Since mid-1999, relations between Turkey and Greece have improved significantly. However, this détente has been limited to relatively non-controversial areas. The core differences over the Aegean and Cyprus have not been resolved.

At the same time, a resolution of the Aegean and Cyprus issues has become increasingly linked to Turkey's quest for EU membership. In accepting Turkey as an official candidate for membership at the Helsinki summit, the EU Council urged candidate states to make every effort to resolve their border disputes and other related issues or, failing that, to bring the dispute to the International Court of Justice (ICJ) in a "reasonable" time. The council stated that it would review the situation at the end of 2004 in relation to the accession process. Thus, if there is no progress toward resolving the Aegean dispute by 2004, the EU could refuse to open accession negotiations with Turkey.

At Helsinki, the EU also agreed that a Cyprus settlement would not be a precondition for Greek Cypriot membership in the EU—a position it reaffirmed in its *Strategy Paper* issued on November 13, 2001. However, it also noted that in reviewing the situation it would take into consideration "all relevant factors." The EU expects to conclude accession negotiations with Nicosia in late 2002. There are few economic obstacles to Nicosia's membership. Thus, in principle, the Greek part of the island could become a member of the EU by 2004–2005.

Greek Cypriot membership would be traumatic for Turkey and would open up the prospect that Greek Cyprus could veto Turkish membership in the EU or demand certain concessions regarding Cyprus as a condition for lifting its veto. A decision by the EU to accept Greek Cyprus before a Cyprus settlement could provoke a crisis in Turkish-EU relations much more profound than the sharp downturn in relations with the EU after the Luxembourg summit. Turkey could respond by integrating the Turkish Republic of Northern Cyprus (TRNC) more closely with the Turkish mainland or by freezing relations with the EU. However, Turkey is unlikely to abandon the goal of membership entirely.

THE DEFENSE AND SECURITY DIMENSION

Turkey's relations with Europe have also been complicated by differences over defense and security policy. Unlike other Southern European members of NATO, Turkey has not witnessed a strong "Europeanization" of its foreign policy. This has tended to highlight Turkey's "distinctiveness" and set it apart from the rest of the southern region in NATO.

The development within the EU of an increasingly important security and defense policy (ESDP) has reinforced this distinctiveness and created a new set of problems. Although other Southern European countries have avidly embraced the EU's development of a stronger security and defense component, Turkey's attitude has been much more ambivalent. Since Turkey is not a member of the EU—and not likely to be one for quite a while—Ankara does not want to see any significant weakening of NATO's role in European security, since this would reduce Turkey's own voice on European security matters.

Turkey is not directly opposed to the expansion of the EU's role in security and defense matters, but Ankara has sought assurances that it will be involved in the planning *and decisionmaking* in EU crisis management operations, especially those that touch directly on its own security interests, and has refused to agree that the EU can draw "automatically" on NATO assets to manage a crisis in which NATO decides not to become involved. Instead, it has insisted that the EU's access to these assets be reviewed on a case-by-case basis.

Turkey's demands for closer consultation and involvement in EU decisionmaking in EU crisis management operations have hindered the coordination of crisis management planning between NATO and the EU.[33]   Ankara's objections have been driven by two main concerns. First, Turkey feared that Greece might use its membership in the EU to push the EU to intervene in areas—particularly Cyprus—that directly affect Turkey's security. Second, most of the crises that the EU or NATO might face in the future are likely to be on or near Turkey's periphery. Thus, Turkey wanted to assure that it would be involved in decisions that directly affected its security.

In early December 2001 Turkey accepted an American/British-sponsored compromise proposal—the "Ankara Document." The Ankara Document provided assurances that the EU's ESDP would not be used against other NATO allies (i.e., Turkey). In addition, it guaranteed that Turkey would be closely consulted in the case of an intervention by the EU's Rapid Reaction Corps in any contingency in the geographic vicinity of Turkey or that affected Turkish security interests.[34]   The Ankara Document thus met Turkey's two main concerns. However, at the EU summit in Laeken, Belgium, a few days later, Greece raised objections to aspects of the Ankara compromise and forestalled its implementation, claiming that the Ankara text would, in effect, give Turkey a veto over Greek national interests such as Cyprus and the Aegean and leave these areas outside the competence of the EU's ESDP. Such a situation is regarded as unacceptable by Athens.

Eventually a compromise is likely to be found that will allow NATO and the EU to coordinate their cooperational plans for crisis management. However, the dispute has served to deepen mistrust between Ankara and Brussels. Many Turks see the dispute as further proof of the EU's desire to exclude Turkey from important decisions affecting its security while Turkish intransigence and brinkmanship on this issue have irritated many EU officials and made them less

---

[33]For background, see Antonio Missiroli, "EU-NATO Cooperation in Crisis Management: No Turkish Delight for ESDP," *Security Dialogue*, Vol. 33, No. 1, March 2002, pp. 9–26.

[34]See Judy Dempsy and Leyla Boulton, "Turkey lifts objections to EU force," *Financial Times*, December 4, 2001. See also Horst Bacia, "Widerspruch zwischen Wollen und Können," *Frankfurter Allgemeine Zeitung*, December 17, 2001. For a detailed discussion, see Missiroli, "EU-NATO Cooperation in Crisis Management: No Turkish Delight for ESDP," pp. 20–23.

inclined to show flexibility on other issues related to Turkey's EU membership.

Turkey has taken a rather cautious approach to NATO's transformation since the end of the Cold War. Ankara initially had reservations about the first round of NATO enlargement, fearing that it would antagonize Russia. However, it has strongly supported the inclusion of Bulgaria and Romania in a second round on the grounds that this would help stabilize the Balkans. At the same time, because Turkey faces a serious threat from Iraq and Syria, Ankara is particularly concerned that NATO's new emphasis on crisis management does not lead to a weakening of the Alliance's commitment to collective defense.

NATO's transformation in the wake of the September 11 attacks could increase Turkey's strategic weight within the Alliance. In the future, NATO is likely to show greater concern for threats beyond Europe's borders. In such a more outward-looking Alliance, Turkey's position on the periphery of the Middle East could enhance Turkey's strategic importance—but also its exposure.[35]

At the same time, Turkey's greater involvement in Middle Eastern affairs since the end of the Cold War creates new dilemmas for Ankara. Many European allies are reluctant to broaden NATO's scope of action beyond Europe and might balk at aiding Turkey if it gets into a conflict with Iraq or Syria, especially if Turkey were perceived to have provoked the conflict.[36] However, a failure to come to Turkey's aid in such a case would create a crisis in Turkey's relations with NATO and might even prompt Turkey to withdraw from the Alliance.

---

[35]The events of September 11 have provoked a debate in Turkey about the effect of the terrorist acts on Turkey's strategic position. For a skeptical view, see Saban Kardas, "The strategic importance of Turkey after September 11," *Turkish Daily News*, May 29, 2002.

[36]Germany's hesitant response to Turkey's request for Allied Mobile Force Reinforcements during the Gulf crisis highlights this problem. To many Germans, deterring a possible attack by Iraq against Turkey was not what NATO was all about. To many Turks, on the other hand, Germany's ambivalent response called into question the validity of Article V (collective defense) of the Washington treaty and raised broader doubts about the utility of NATO membership. See Ian O. Lesser, *Bridge or Barrier: Turkey and the West After the Cold War*, pp. 14–15.

This issue could become more acute if the United States launches an attack against Iraq. Turkey has expressed reservations about such an attack, which would increase Ankara's own exposure, especially if it allowed the United States to use its facilities to conduct strikes against Iraq. However, most Turks believe that Turkey will have little choice but to support the United States if Washington eventually decides to attack Iraq.

Turkey's proximity to the Middle East also gives it a special interest in counter-proliferation and ballistic missile defense. Turkey is the only NATO member that currently faces a threat from ballistic missiles launched from the Middle East (although the threat to other Southern European countries will increase over the next decade).[37] Thus, as the ballistic missile threat intensifies and the United States proceeds with the construction of its missile defense system, Turkey is likely to show increased interest in developing a regional missile defense architecture in cooperation with the United States and Israel and possibly even India.

THE AMERICAN FACTOR

The events of September 11 are also likely to have an important effect on Turkey's relationship with the United States—and indirectly the EU. Turkey has always seen a strong tie to Washington as an important component of its security. The United States, in turn, has been one of Turkey's strongest allies and has been more supportive than many EU members of key Turkish priorities such as the construction of the Baku-Ceyhan pipeline, the campaign against PKK terrorism, and Turkey's quest for EU membership.

The United States has strongly supported Turkey's candidacy for EU membership, largely for strategic reasons. In the past, U.S. "lobbying" for Turkey's candidacy often was a source of friction in U.S.-European relations.[38] However, with the Helsinki decision, the

---

[37]See Ian O. Lesser and Ashley J. Tellis, *Southern Exposure: Proliferation Around the Mediterranean*, Santa Monica, CA: RAND, 1996.

[38]For details, see F. Stephen Larrabee, "U.S. and European Policy Toward Turkey and the Caspian Basin," in Robert D. Blackwill and Michael Stürmer, eds., *Allies Divided: Transatlantic Policies for the Greater Middle East*, Cambridge, MA: The MIT Press, 1997, pp. 143–173.

United States essentially achieved its main objective—to ensure that Turkey was accepted as a genuine candidate for EU membership— and tensions between the United States and its European allies over Turkey largely subsided, especially since it was evident that Turkey still had a long way to go before it would actually be ready to join the EU.

The events of September 11, however, have served to strengthen the U.S.-Turkish strategic partnership and could cast the issue of Turkey's membership in the EU in a new light. Washington sees Ankara as a critical ally in the war against terrorism. Thus, for strategic reasons it may be more inclined in the future to side with Turkey in disputes with the EU. This, in turn, could lead to the re-emergence of the tensions that characterized U.S.-EU relations in the period leading up to Helsinki.

THE DIFFICULT PATH AHEAD

Turkey today stands at a historic crossroads. The Helsinki summit has opened the possibility of EU membership. But membership will require extensive changes in the Kemalist system that has evolved over the last 78 years, especially the reduction in the role of the military in guiding and directing Turkey's political evolution. It will also require the Turkish elite to accept a greater degree of political and social pluralism as well as unprecedented limits on state sovereignty. These changes are bound to provoke resistance from forces that have a strong vested stake in the maintenance of the current system. But they cannot be avoided if Turkey hopes to become a full member of the EU.

The reform package adopted by the Turkish parliament on August 3, 2002, goes a long way to meeting many of the Copenhagen criteria, especially regarding the abolition of the death penalty and Kurdish broadcasting and education.[39] The ball is now in the EU's court. But the EU is likely to want to see how well the laws are actually imple-

---

[39]For a detailed discussion of the reform package, see Heinz Kramer, "Ein wichtiger Schritt in Richtung EU," SWP-Aktuell 29, Ebenhausen: Stiftung Wissenschaft und Politik, August 2002.

mented before it is willing to set a date for opening accession negotiations.

The victory by the Islamist Justice and Development Party (AKP) in the November 3, 2002, parliamentary elections is likely to reinforce this caution. Although the AKP supports Turkey's entry into the EU—in large part because it sees EU membership as a constraint on any possible crackdown on its existence by the military—the EU is likely to want to see how the AKP performs in office before setting a date for opening accession negotiations.

Developments within the EU will also have an effect on Turkey's aspirations for membership.

Developments within the EU will also be important. A slowdown in the EU integration process could negatively affect Turkey and diminish its chances for membership over the long run. In many parts of Europe, support for EU enlargement has been declining, as the costs of enlargement have become clearer. The strong showing by the far right in elections in Austria, Italy, France, Denmark, and Holland could strengthen this trend and complicate Turkish aspirations for membership. A failure by the EU to offer Turkey a date for opening accession negotiations—or at least a clear perspective for opening negotiations—at its summit in Copenhagen in December 2002 could also dim Turkish aspirations and provoke a strong domestic backlash in Turkey similar to the one that occurred in 1997 after the Luxembourg summit.

Chapter Four

RELATIONS WITH GREECE AND THE BALKANS

Turkey's relations with Greece form an important part of Turkey's broader agenda. The conflict between the two countries has been a persistent threat to security in the Eastern Mediterranean since the mid-1950s. During the Cold War, the differences between the two countries threatened to break out into open conflict on several occasions.[1] However, these differences have taken on added importance since the end of the Cold War for several reasons.

First, the Aegean has been one of Europe's most dangerous flash-points. Turkey and Greece have come close to armed conflict several times in the last two decades—most recently in January–February 1996 over the islets of Imia/Kardak. Only last-minute U.S. diplomatic intervention prevented an armed confrontation. Although relations have improved since mid-1999, as long as the issues that gave rise to the near clash remain unresolved, there is always a danger that an incident could lead to inadvertent armed conflict.

Second, the Cyprus issue continues to aggravate Greek-Turkish relations. Since 1974, the division of the island has hardened, with little communication and interaction between the Turkish and Greek Cypriot communities. In addition, the island has become increasingly militarized. In recent years, both sides have sought to augment

---

[1]For background, see Theodore A. Couloumbis, *The United States, Greece and Turkey: The Troubled Triangle*, New York: Praeger, 1983; Tozun Bahcheli, *Greek-Turkish Relations Since 1955*, Boulder, CO: Westview Press, 1990; Monteagle Stearns, *Entangled Allies: U.S. Policy Toward Greece, Turkey and Cyprus*, New York: The Council on Foreign Relations, 1992; and James Brown, *Delicately Poised Allies: Greece and Turkey*, London: Brassey's, 1991.

their military capabilities on the island. This growing militarization has increased the dangers of a Turkish-Greek confrontation, as the crisis over the S-300 missiles in 1997–1998 highlights.

Moreover, the lack of a settlement of the disputes over the Aegean and Cyprus is an obstacle to Turkey's relations with the EU. At the Helsinki summit in December 1999, the EU accepted Turkey as a candidate member, but it made a resolution of Turkey's differences with Greece over the Aegean and Cyprus a prerequisite for Turkey's eventual membership in the EU. Thus, Turkey's European aspirations are now directly tied to a resolution of its differences with Greece over the Aegean and Cyprus.

Moreover, the new détente is fragile and by no means irreversible. The core differences between the two countries—the Aegean and Cyprus—have yet to be seriously addressed. In addition, a number of issues, particularly the EU's decision regarding the Greek Cypriot application for membership, could have a significant effect on Greek-Turkish relations, undermining the recent improvement in bilateral ties and possibly even setting the stage for a new period of confrontation.

THE HISTORICAL LEGACY

The current difficulties in Turkish-Greek relations have deep historical roots that directly affect how each side perceives and relates to the other. These roots touch directly on important issues of national identity.[2] The modern Greek state was born of a struggle against Ottoman rule and for much of the next 80 years it expanded by lopping off parts of Ottoman territory. The *Megali Idea*—the desire to unite Greeks in a common Greek state—was a driving force behind Greek policy up until the early 20th century. Thus, Greece's own independence and process of state-building have been closely tied to a struggle against Turkey.

Similarly, the birth of the modern Turkish state was closely associated with the War of Independence and the campaign against Greece

---

[2]See Heinz Kramer, "Turkey's Relations with Greece: Motives and Interests," in Dimitri Constas, ed., *The Greek-Turkish Conflict in the 1990s,* New York: St. Martin's Press, 1991, pp. 57–72.

that ended with Atatürk's expulsion of the Greek forces from Central and Western Anatolia in 1922 and the subsequent exchange of populations in 1923–1924.[3] Thus, both states link their existence and an important part of their identity to experiences that are associated with negative images of the other side. This has served to reinforce a sense of mutual mistrust that has complicated the resolution of outstanding differences.

On rare occasions, however, Turkey and Greece have shown an ability to put aside their differences and cooperate. The most important example of this capacity occurred during the interwar period. Only eight years after Greece's defeat at the hands of Turkey, Atatürk and Greek Prime Minister Eleftherios Venizelos signed the Ankara Agreement, resolving outstanding issues left over from the earlier confrontation, such as property rights arising out of the exchange of populations. This was followed by the conclusion of a formal Friendship and Cooperation Treaty in 1933, which introduced a period of cordiality and cooperation that lasted into the early postwar period.

The onset of the Cold War and the threat posed by the Soviet Union initially created strong incentives for both countries to put aside their differences. But beginning in the mid-1950s, relations became increasingly strained over the Cyprus issue. The attempted coup against President Makarios of Cyprus by the Greek junta in July 1974 and the subsequent Turkish invasion of the island marked an important turning point in relations. Thereafter, relations remained, until very recently, tense.

In the past two decades, there have been several attempts to overcome these difficulties and improve relations. The most important was the meeting between Turkish Prime Minister Turgut Özal and Greek Prime Minister Andreas Papandreou in Davos in January 1988.[4] The two leaders agreed to establish a hotline, to meet at least

---

[3]For a comprehensive discussion, see Harry J. Psomiades, *The Eastern Question: The Last Phase*, New York: Pella, 2000.

[4]For a good discussion of the Davos process, see Richard Clogg, "Greek-Turkish Relations in the Post-1974 Period," in Dimitri Constas, ed., *The Greek-Turkish Conflict in the 1990s*, New York: St. Martin's Press, 1991, pp. 12–23, and John Koliopoulos and Thanos Veremis, *Greece, the Modern Sequel*, New York: New York University Press, 2002, pp. 307-314. For a Turkish perspective, see Mehmet Ali Birand, "Turkey and the Davos Process," in Constas, *The Greek-Turkish Conflict in the 1990s*, pp. 27–39.

once a year, and to visit each other's country. They also called for an intensification of contacts. In an early gesture of goodwill, Turkey rescinded the 1964 decree restricting the property rights of Greek nationals in Turkey.

These moves led to a thaw in bilateral relations (the Davos Process). But the thaw proved short-lived because it lacked strong domestic support. Neither leader sought to build bureaucratic and public support for the attempt at reconciliation. Thus, both leaders found it difficult to "sell" the thaw to a skeptical domestic audience at home, especially as their own internal positions weakened. As a result, the détente gradually lost momentum and eventually collapsed.

THE AEGEAN DISPUTE

There are three main sources of tension between Turkey and Greece. The most important of these centers around the Aegean. The Aegean dispute is not really one dispute but a series of disputes: These relate to limits on territorial sea, sovereign rights over the continental shelf and airspace, management of the military and civil air-traffic control zone, and the militarization of the Greek islands.[5]

For Turkey, the most important of these issues is the territorial sea issue. Under the 1982 Law of the Sea Convention—which Turkey has not signed—Greece has the right to extend its territorial waters to 12 miles although it has so far refrained from exercising that right. Greek extension of the territorial waters from six to 12 miles would make Turkish access to major ports, such as Istanbul and Izmir, more difficult. Turkey has repeatedly said that any effort by Greece to extend its territorial waters to 12 miles would constitute a *casus belli*. This explicit threat to use force if Greece exercised its right to extend its territorial waters has been a prime source of tension between the two countries. Greek officials insist that there can be no far-reaching rapprochement between Greece and Turkey as long as Turkey continues to threaten to use force to settle outstanding

---

[5]For a detailed discussion of the Aegean dispute, see Andrew Wilson, *The Aegean Dispute*, London: International Institute for Strategic Studies, 1980. Also Heinz-Jürgen Axt, "Der Ägäis Streit—ein unlösbarer griechisch-türkischer Konflikt?" *Südosteuropa Mitteilungen*, Nr. 2, 1999, pp. 137–151.

issues, especially ones in which Greek actions are consistent with international law.

Turkey and Greece are also at odds over the Aegean airspace. Turkey rejects the 10-mile airspace claimed by Greece, arguing that Greece is entitled to exercise sovereignty only over six miles. To emphasize this point, Turkey regularly sends its aircraft up to six miles from the Greek coast. Greece responds to what it considers violations of its airspace by sending aircraft to intercept the Turkish aircraft. The mock dogfights and aerial challenges are a source of constant concern to Turkey's NATO allies, who worry that some incident or miscalculation could lead to a major confrontation between the two countries, as nearly happened during the Imia/Kardak crisis in January/February 1996.

Turkey has offered to open a bilateral dialogue with Greece to resolve these issues. However, Greece has rejected a broad-based bilateral dialogue, arguing that there is only one issue that needs to be resolved—the continental shelf. This issue, Greece argues, should be submitted to the ICJ at the Hague for adjudication. However, Turkey has refused to submit the issue to the ICJ, preferring instead to resolve the issue through bilateral negotiations, where it feels it has more leverage.

Another issue burdening bilateral relations is Greece's militarization of the Eastern Aegean and Dodecanese islands, which Turkey argues violates the Treaty of Paris (1947). Greece argues that its militarization of the islands is a defensive response to the creation by Turkey of the 100,000-man strong "Fourth Aegean Army," which was formed shortly after the Turkish invasion of Cyprus in 1974 and which has its headquarters in Izmir, just a few miles from the Greek islands. This army, Greece maintains, poses a serious threat to Greek sovereignty.

So far, Turkey has refused to disband the Aegean army. However, in the spring of 2000, Admiral (Ret.) Güven Erkaya, the former commander of the Turkish navy and an advisor to Prime Minister Ecevit, proposed in a secret memorandum (later leaked to the Turkish press) that Turkey disband the Aegean army in return for a pledge by

Greece not to extend its territorial waters to 12 miles.[6] Although his proposal was rejected by the Turkish military,[7] who maintained that Greece had to first agree not to extend its territorial waters to 12 miles, it continues to be seen in some Turkish circles as a possible avenue that could, if combined with reciprocal measures by Greece, lead to a de-escalation of the Aegean dispute.

Turkey's relations with Greece have significantly improved since mid-1999 (see below), but the dispute over the Aegean continues to cast clouds over the relationship and could even derail the recent détente. In October 2000, for instance, differences over the inclusion of the islands of Lemnos and Ikaria prompted Greece to withdraw from NATO exercises in the Aegean, casting a pall over the rapprochement between Greece and Turkey.[8] Although both sides have continued to stress their commitment to improving relations, the incident underscores the degree to which the differences over the Aegean continue to burden the relationship.

In addition, the EU has made a resolution of Turkey's differences with Greece a requirement for Turkish accession to the EU. The communiqué issued at the EU's summit in Helsinki in December 1999 urged candidate states to make every effort to resolve their border disputes and other related issues or, failing that, to bring the dispute to the ICJ in a reasonable time. The council stated that it would review the situation by the end of 2004 in relation to the accession process. Thus, Turkey's application for EU membership could be held up if there is no resolution of the dispute.

In January 2002, Greece agreed to open a bilateral dialogue with Turkey on Aegean issues. Greek diplomats want the dialogue limited to the issue of the continental shelf. However, in the course of the dialogue other issues, such as differences over airspace control and

---

[6]Leyla Boulton, "Turkey signals softer line in Greek links," *Financial Times*, May 24, 2000. For the text of the memorandum, see Ali Ekber Ertürk, "Aegean Army Should Be Abolished," *Sabah*, May 22, 2000, translated in FBIS-WEU-2000-0524, May 22, 2000.

[7]"Dismantling Aegean army not on agenda," *Turkish Daily News*, May 24, 2000. See also "Turkish officials: before we dismantle the Aegean army, Greece has to recognize six mile limit," ibid., May 23, 2000.

[8]For background, see John Ward Anderson, "Flap During NATO Drill Upsets Greek-Turkish Thaw," *The Washington Post*, October 26, 2000. Also, "Aegean Rumpus," *The Economist*, October 28, 2000.

the Ecumenical Patriarch (the spiritual leader of the world's Orthodox Christians, who resides in Istanbul), may also be discussed.

THE CYPRUS CONFLICT

Cyprus is a second major source of tension between Turkey and Greece. During the 1930s and 1940s when Cyprus was under British rule, the fate of the Turkish Cypriots was not a burning issue for the Turkish government. It became a major concern only in the 1950s when the Greek Cypriots, supported by the Greek government, intensified their demands for *enosis* (union with Greece) and the British government began considering relinquishing control over the island. Although Turkey preferred a partition of the island (*taksim*), it agreed to independence on the condition that the Turkish Cypriots would have the status of a community with equal rights with the Greek community.

From the Turkish point of view, independence was a second-best solution. It was accepted because it prevented *enosis* and provided important legal guarantees of equality for the Turkish Cypriot community.[9] Under the 1960 Treaty of Guarantee, Turkey became one of the three guarantor powers (along with Greece and Britain) of the island's independence. This ensured Turkey a certain degree of influence over developments on the island and gave Ankara the right to intervene, either singularly or collectively, to reestablish the constitutional arrangements on the island if they were violated. It was under the provisions of the 1960 Treaty of Guarantee that Turkey justified its invasion of the island in 1974.

The 1960 constitutional arrangements, however, proved unworkable and collapsed when President Makarios sought to amend the Constitution.[10] His changes would have relegated the Turkish Cypriots to the status of a minority rather than a community with equal political

---

[9]See Clement H. Dodd, "A Historical Overview," in Clement H. Dodd, *Cyprus: The Need for New Perspectives* , Huntingdon, England: The Eothen Press, 1999, pp. 1–15. See also his *The Cyprus Imbroglio*, Huntingdon, England: The Eothen Press, 1998.

[10]For a detailed discussion, see Dodd, *The Cyprus Imbroglio*; Nancy Crawshaw, *The Cyprus Revolt: An Account of the Struggle for Union with Greece*, London: George Allen & Unwin, 1978; and Polyvios Polyviou, *Cyprus: Conflict and Negotiations, 1960–1980*, London: Duckworth, 1980.

rights with the Greek community. This was unacceptable to both the Turkish Cypriots and Turkey. When anti-Turkish violence broke out in 1964, Turkey threatened to intervene. However, Ankara was deterred from carrying out the threat by President Johnson's blunt letter to Turkish Prime Minister Ismet Inönü warning that the United States and NATO could not guarantee Turkey's protection if a Turkish invasion provoked Soviet intervention.[11]

Turkey showed only lukewarm support for the Turkish Cypriots in the period 1965–1973. However, Turkish policy hardened after the election of Bülent Ecevit as prime minister at the end of 1973. When the military junta in Athens sought to overthrow Makarios and install a pro-*enosis* extremist, Nicos Samson, as president, Ecevit ordered the invasion of the island. However, Turkey invaded only after first requesting that Britain participate in the invasion under the terms of the 1960 Treaty of Guarantee. When Britain refused to participate, Turkey decided to act unilaterally.

The 1974 invasion set the contours for the current conflict. It led to the expulsion of 200,000 Greek Cypriots from their homes and the division of the island into two autonomous administrations, one Greek Cypriot, the other Turkish Cypriot. Since then, Turkey has maintained 30,000 to 35,000 troops on the island. The Greek Cypriots regard the troops as occupation troops. Turkey, however, see the troops as the main guarantee of the rights of the Turkish Cypriots.

From the Turkish perspective, the invasion "solved" the Cyprus problem. In 1983, the Turkish Cypriot part of the island declared itself an independent state—the TRNC—with Rauf Denktash as its president. The TRNC is recognized only by Turkey and is economically and politically dependent upon Ankara, which heavily subsidizes its economy. Although this subsidy imposes a substantial economic burden on Turkey, Ankara has considered the political and strategic benefits worth the economic costs.

---

[11]For the text of the Johnson letter, see the *Middle East Journal,* Vol. 20, Summer, 1966, pp. 386–393. For a detailed discussion of the effect of the letter on U.S.-Turkish relations, see George H. Harris, *Troubled Alliance: Turkish-American Problems in Historical Perspective,* Washington, D.C.: American Enterprise Institute, 1972, pp. 105–124.

Turkey's approach to Cyprus has undergone an important shift since the mid-1990s. Before 1997, Turkey put primary emphasis in its Cyprus policy on the protection of the Turkish Cypriot community on the island. Since 1997, however, Turkey has increasingly emphasized that Cyprus is a security issue for Turkey in its own right. Prime Minister Bülent Ecevit in particular is a firm proponent of this view. For years Ecevit argued that Cyprus should be seen not as a burden but as an important component of Turkey's own security.[12]

Ankara sees Cyprus as a cornerstone of Turkish security and a key element of the defense of Anatolia. Cyprus is also increasingly seen as important for the protection of Caspian oil that is expected to flow through the Baku-Ceyhan pipeline and then be transported on to Europe. Thus, Turkey wants to ensure that the island is not controlled by a hostile power, especially Greece. Many Turks believe that the best way to ensure that this does not happen is to keep the island divided and maintain a strong military presence there.

Turkey has reacted harshly to efforts to change the military status quo such as the Greek Cypriot decision to acquire S-300 missiles from Russia. Ankara threatened to use military force, if necessary, to prevent the deployment of the missiles on Cyprus.[13] The crisis was finally defused when the Greek Cypriot government—under strong U.S. and EU pressure—agreed to deploy the missiles on Crete rather than the Cypriot mainland. However, the bellicose Turkish response to the initial threat to deploy the missiles on the Cypriot mainland highlights the importance that Turkey attaches to maintaining the political and military status quo on the island.

Since the late 1990s, Turkey has taken steps to strengthen its ties with the TRNC. In July 1997, Turkey and the TRNC issued a joint statement expressing their determination to strengthen and deepen cooperation. The declaration called *inter alia* for the establishment of

---

[12]As he stressed at the celebration of the 25th anniversary of the Turkish invasion of Cyprus in July 1999: "As much as Turkey is the generator of KKTC [TRNC] security, the KKTC is the guarantee of Turkey's security." See "Turkey and Cyprus Not Moving An Inch From Cyprus Policy," *Turkish Probe*, July 25, 1999.

[13]For a detailed discussion, see Madeleine Demetriou, "On the Long Road to Europe and the Short Path to War: Issue-Linkage Politics and the Arms Build-Up in Cyprus," *Mediterranean Politics*, Vol. 3, No. 3, Winter 1998, pp. 38–51.

an Association Council, the creation of an economic and financial union between the two states, the inclusion of the TRNC in Turkey's regional development master plan, and the improvement of transportation links between Turkey and the TRNC.[14] Turkey also began to include Turkish Cypriots in Turkish delegations and missions abroad. At the same time, Turkey moved away from the bi-zonal and bi-communal federation that it supported for nine years, insisting on the recognition of two separate states as a basis for any solution.

The passage of time, moreover, has led to a hardening of the status quo. At present, a whole generation of Turkish and Greek Cypriots has grown up with virtually no contact with one another. This situation is likely to grow worse with time, leading to even greater estrangement and isolation between the two communities. In addition, the ethnic composition of the TRNC is changing. As a result of the massive influx of new settlers from the Turkish mainland and the emigration of educated Turkish Cypriots, the proportion of Turkish Cypriots in the TRNC is declining. Today, Turkish Cypriots make up only about 60 percent of the TRNC's population. If the economic situation in the TRNC continues to deteriorate, their number is likely to further decline, as more and more Turkish Cypriots—especially educated ones—emigrate because they cannot find jobs.[15]

PROSPECTS FOR A SETTLEMENT

Intercommunal talks between the two Cypriot communities have been conducted, under UN auspices, since 1974. However, they have produced no major breakthrough. The basic problem is that the two Cypriot sides have very different visions of the island's political future. The Greek Cypriots want a bi-zonal, bi-communal federation with broad powers for the central government. Since 1998, the Turkish Cypriots, by contrast, have pressed for a

---

[14]For the text of the Joint Statement, see Dodd, *The Cyprus Imbroglio*, pp. 190–192 (Appendix 9).

[15]Since 1974 an estimated 50,000–60,000 Turkish Cypriots—more than one-third of the Turkish Cypriot population—have emigrated. See "Ankara's Zypern–Politik in der Sackgasse?" *Neue Zürcher Zeitung*, May 23, 2001.

confederation with weak federal powers, based on the recognition of two separate and equal states. These two visions are in direct conflict with one another and are difficult to reconcile.

Prospects for a settlement before the EU summit in Copenhagen in December 2002 appear bleak. On January 16, 2002, the leaders of the two Cypriot communities, Rauf Denktash and Glafkos Clerides, agreed to open negotiations, with the goal of achieving an agreement on the island's future by the summer of 2002. However, these talks have made little progress. There have been some minor changes in approach on both sides, but neither side has substantially altered its fundamental position.

The Greek Cypriot side seems to be playing for time in the belief that Greek Cyprus will be invited to join the EU, with or without a settlement. Thus, it has little incentive to make any major concessions. The Greek Cypriots appear to calculate that their leverage will increase once they are a member of the EU and that EU membership will make a settlement of the conflict on their own terms easier. Hence, they seem content to "play out the EU clock." At the same time, the Turkish Cypriot side also appears unwilling to depart from its insistence on a two-state solution.

Ultimately, the key to a Cyprus settlement lies in Ankara. However, there is little likelihood of a shift in Turkish policy—if at all—until after the national elections in November 2002. By then, however, it may be too late. It seems almost certain that the Greek part of the island will be invited to join the EU at the EU summit in Copenhagen in December 2002.

Greek Cyprus membership would be a real trauma for Turkey and could lead to a serious deterioration of Turkey's relations with the EU. Turkey's chances of opening accession negotiations with the EU would be jeopardized and prospects for obtaining EU membership in the foreseeable future would be seriously set back. It could also have a spillover effect on Turkey's relations with Greece, endangering the current bilateral détente. Indeed, a new period of Greek-Turkish tension might well ensue.

THE MINORITY ISSUE

A third irritant in Turkey's relations with Greece has been the status and treatment of the Turkish minority in Greece. There are about 120,000 Muslims in Greece, the majority of whom are ethnic Turks. The rest are mostly Gypsies (Roma) and Pomaks (Muslims with an affinity to Bulgarian culture).[16] Until recently, Greece has insisted on using the term "Muslim" for this minority, even though a large part of the Muslim population is composed of ethnic Turks. Greece bases its position on the Lausanne Treaty (1923) which refers to the population as "Muslims."[17]

The Muslim/Turkish minority in Greece basically enjoys the same rights as ethnic Greek citizens of Greece. However, the minority has been subject to indirect forms of discrimination regarding the purchase of land, obtaining building permits for the construction of private buildings and mosques, obtaining driving licenses, and having their land expropriated for public use. Until 1998, the minority was also subject to possible loss of Greek citizenship if members of the minority left Greece.

The plight of the Turkish minority receives considerable attention in the Turkish media, which claims that the community faces systematic discrimination. Greece, in turn, accuses the Turkish government of having systematically forced out the Greek minority in Istanbul. In 1923, the Greek population in Istanbul numbered about 120,000. Today it has dropped to less than 3,500. Most of the Greek minority left in the 1950s and 1960s when Greek-Turkish relations were tense as a result of the Cyprus crisis.

Greece denies that it discriminates against the Turkish minority and in recent years the Greek government has undertaken a number of steps to improve the lot of the Turkish minority. In 1998, the government revised the controversial provisions of the Greek Nationality Law that had been used to revoke the citizenship of members of the

---

[16]For a detailed discussion, see Hugh Poulton, *The Balkans: Minorities and States in Conflict*, London: Minority Rights Publications, 1991, pp. 182–188; and Tozun Bahcheli, *Greek-Turkish Relations Since 1955*, pp. 169–187.

[17]However, in the 1940s and 1950s, both terms "Muslim" and "Turk" were used to refer to the population. Greece reverted to using solely the term "Muslim" only after the tensions with Turkey increased in the 1960s.

Turkish minority who traveled abroad. Greece has also taken measures to develop the economy of Western Thrace, where most of the Turkish minority lives.

Recently, moreover, Greece has begun to adopt a more open approach regarding the existence of the Turkish minority. In July 1999, Greek Foreign Minister George Papandreou suggested in an interview that Muslims who felt themselves to be Turks should be allowed to call themselves Turks.[18] His remarks were welcomed in Turkey as a sign of Athens' desire to improve relations with Ankara. Although they caused a storm of protest in nationalist circles in Greece, since then there has been a growing acceptance in Greece of the idea that there is a Turkish minority.

These changes have helped to defuse the minority issue as a source of tension in bilateral relations. However, the leaders of the Turkish minority and the Turkish government continue to press Greece to take further measures to improve the minority's economic status and educational opportunities. In particular, they want the minority to be able to elect its own religious leaders (*Muftis*) rather than have them appointed by the Greek government, as is currently the case.

NEW REGIONAL GEOMETRIES

The conflict between Turkey and Greece has also been given sharper focus by the emergence of new regional alliances. Just as Turkey has expanded its relations in the Balkans, Greece has sought to cultivate new strategic allies in the Caucasus and the Middle East. Greece's effort to forge closer ties to Armenia has aroused particular suspicion in Ankara. Ankara regarded the Greek-Armenian defense agreement signed in 1996 as specifically directed against Turkey.

More upsetting, however, was Greece's effort to forge closer ties— especially defense ties—to Syria because of Syria's support (until 1998) of the separatist Kurdistan Workers Party (PKK) and its leader Abdullah Öcalan, regarded in Turkey as a terrorist. Reports that Greece had signed a substantive "defense agreement" with Syria in

---

[18]See "Griechisch-türkischer Dialog mit Begleitmusik," *Neue Zürcher Zeitung,* July 31, 1999. Also "Georges Papandréou s'attaque aux préjugés anti-turcs," *Le Monde*, October 13, 1999.

1995 appear to be exaggerated. But the closer cooperation between Greece and Syria caused considerable concern in Ankara because of the PKK connection and Turkish complaints that Greece was supporting PKK activities on its soil.[19]

Turkish suspicions were reinforced by revelations that Öcalan had been smuggled into Greece and was given sanctuary in the Greek embassy in Nairobi. Possibly tipped off by U.S. intelligence, Turkish authorities managed to capture Öcalan as he was being whisked off to the Nairobi airport. The whole affair was a major embarrassment for the Greek government and led to a sharp deterioration of Greek-Turkish relations as well as the dismissal of several high-ranking Greek officials, including then Foreign Minister Theodore Pangelos.[20]

EARTHQUAKE DIPLOMACY AND THE NEW DETENTE

Although the Öcalan affair led to a sharp deterioration in Turkish-Greek relations, paradoxically, it also served as an important stimulus to an eventual thaw in relations. In the aftermath of the Öcalan affair, both sides began a quiet dialogue designed to explore ways to improve relations. This dialogue was given important momentum by the devastating earthquake in Turkey in August 1999 and the much smaller one in Athens several weeks later. The rapid and generous support by Greece to the Turkish earthquake victims had an important psychological effect on the Turkish public. In the wake of the earthquake, each side began to see the other in human terms rather than as an abstract enemy. This helped to break down old stereo-

---

[19]While the Greek government repeatedly denied Turkish charges regarding support for the PKK, a number of Greek parliamentarians maintained contacts with the PKK. In January 2000, for instance, Deputy Parliamentary Chairman Panayiotis Sghouridhis, PASOK (Pan-Hellenic Socialist Movement) deputy from Xanthi in Thrace, revealed that he had met with Öcalan in Damascus in 1995, as part of a Greek parliamentary delegation, and again in Rome, in December 1998. Sghouridhis said that he had sent memoranda about these meetings and other contacts with the PKK to Foreign Minister Theodore Pangelos and Parliamentary Chairman Apostolos Kaklamanis. See Ionnis Dhiakoyannis, "The Secret Meeting with Öcalan in Rome," *Ta Nea*, January 24, 2000, translated in FBIS-WEU-2000-0125, January 24, 2000.

[20]For a detailed discussion, see in particular Gülistan Gürbey, "Der Fall Öcalan und die türkisch-griechische Krise: Alte Drohungen oder neue Eskalation?" *Südosteuropa Mitteilungen*, Nr. 2, 1999, pp. 122–136.

types. At the same time, it provided domestic cover for diplomatic initiatives on both sides and helped to insulate them from strong domestic criticism.

Since the EU summit in Helsinki, the thaw has gained new momentum. In January 2000, Greek Foreign Minister George Papandreou paid a visit to Ankara where he signed five low-level agreements on issues of environment, terrorism, illegal immigration, etc. Papandreou was the highest-ranking Greek official to pay a state visit to Ankara in 38 years. The following month, Turkish Foreign Minister Ismael Cem visited Athens—the first visit by a Turkish foreign minister to Greece in 40 years. Greece and Turkey have also begun a dialogue on confidence-building measures.

These moves have been followed by other important steps to improve relations, including a commitment to remove mines along the Greek-Turkish border, plans to extend the Ignatia highway from Western Greece to Istanbul, an agreement on cooperation on transporting Caspian and Egyptian gas through Turkey and Greece and on to the rest of Europe, joint investment on tourism related to the 2004 Olympic Games in Athens, an agreement on cooperation in dealing with natural disasters, and an agreement on the repatriation of illegal immigrants. The latter agreement is considered to be particularly important because both countries, especially Greece, have been inundated with a large influx of illegal immigrants in recent years.

Important steps have been taken in the defense field as well. In April 2001, Greece announced changes in its military doctrine, ending the state of war mobilization with Turkey that had existed since the 1974 Turkish invasion of Cyprus. In addition, Athens announced plans to cut arms procurement by $4.4 billion, including postponing the purchase of 60 Euro-fighters until 2004, and to reduce its armed forces from 140,000 to 80,000–90,000 men.[21] In April 2000, Turkey also decided to postpone defense spending by $19.5 billion. In both cases these measures were primarily dictated by economic considerations, but they contributed to improving the overall climate of bilateral relations.

---

[21]Andrew Borowiec, "Greek Cutbacks Ease Military Tensions," *The Washington Times*, April 10, 2001.

Energy has also emerged as an important area of bilateral cooperation. In March 2002, the two countries signed a $300 million deal to
extend an Iranian natural gas pipeline from Turkey to Greece. The
pipeline, due to be completed in 2005, could be extended to Italy
with financial assistance from the EU and would help Turkey and
Greece enter the European market both as buyers and sellers.

The key question is whether the current thaw represents a strategic
shift in relations or just a tactical thaw. There have been a number of
efforts to mend fences before—the most notable being the effort by
Özal and Andreas Papandreou after their meeting in Davos in January 1988. All were short-lived and ultimately collapsed. However,
the current effort at détente is likely to prove more enduring than
previous efforts for several reasons.

First, the new détente has strong domestic support. The thaw following the Davos meeting between Papandreou and Özal represented
an attempt to break the logjam in bilateral relations through
"personal diplomacy." But, as noted above, it lacked a solid institutional base and strong domestic support. Thus, it soon collapsed.
The current rapprochement, by contrast, has a much stronger domestic base in each country. The earthquakes in Turkey (August
1999) and Greece (September 1999) created a kind of "bonding" at
the popular level that was absent at the time of the Papandreou-Özal
dialogue. There has also been an effort to involve civic groups in the
rapprochement process, especially the business community.

Second, the current rapprochement represents a strategic shift in
Greece's approach to Turkey. For years, Greece sought to use its
membership in the EU to isolate Turkey in an effort to force Turkey
to change its approach to the Aegean and Cyprus. In the past,
Greece, persistently blocked the dispersal of EU financial aid to
Turkey, linking it to Turkey's policies on the Aegean and Cyprus. It
also blocked Turkey's candidacy for EU membership on the same
grounds.

Greece has now abandoned this approach. Instead, it has adopted a
policy of engagement with Turkey, which is based on the premise
that a more "Europeanized" Turkey is in Greece's long-term interest.
At the EU's Helsinki summit in December 1999, Greece lifted its veto
against Turkey's EU candidacy. This removed the most important

obstacle to the EU's acceptance of Turkey's EU candidacy at the summit (although there is still considerable skepticism in Germany and elsewhere in Western Europe about putative membership).

Third, the détente has been buttressed by growing economic cooperation, especially in the energy field. This cooperation ties the two countries together economically and gives the cooperation a strong economic component. Each side now has a strong economic stake in continuing the détente and would suffer adverse economic consequences if the détente were to collapse. This gives both sides a strong incentive to keep the current rapprochement on track.

Finally, the EU has shifted its approach to Turkey. Before the EU summit in Helsinki, the EU had refused to accept Turkey as a candidate for EU membership—a position that had led to serious strains in Turkey's relations with the EU. However, at Helsinki, the EU Council officially accepted Turkey as a candidate member, opening up the possibility of Turkish membership over the long run. At the same time, the EU made clear that a resolution of Turkey's differences with Greece over the Aegean and Cyprus were a precondition for membership. Thus, Turkey now has a strong incentive to regulate its relations with Greece.

These developments have changed the context of relations and improved the prospects for a far-reaching détente between the two countries. However, the current rapprochement remains fragile for several reasons.

First, most of the changes have come on the Greek side. Without some reciprocal gestures on Turkey's part, it may prove difficult to maintain domestic support in Greece for the rapprochement over the long run. At the moment, most Greeks are willing to give Prime Minister Costas Simitis and Foreign Minister George Papandreou—the chief architect of the recent détente—the benefit of the doubt. But, at some point they may begin to ask what Greece has received in return. Thus, some reciprocal gestures by Turkey will be important to keep the process moving.

Second, so far the rapprochement has been limited mainly to noncontroversial areas such as trade, the environment, and tourism, although new protocols on combating international crime and terrorism have brought cooperation to more difficult issues. But, at some point, the sensitive issues in the Aegean and Cyprus will have to be addressed if the rapprochement is to prove durable.

Finally, the structure of post-Helsinki rapprochement between Greece and Turkey is heavily dependent on the course of Turkey's relations with the EU. If Turkey's candidacy proves hollow or Turkey's political evolution makes integration difficult, Greek-Turkish relations could suffer and the assumptions on which the détente has rested could be undermined. This could lead to a new period of antagonism and confrontation.

THE AMERICAN FACTOR

Turkey's relations with Greece have been—and will continue to be—significantly influenced by U.S. policy. In general, the United States has tried to avoid taking sides in the dispute between Greece and Turkey and to act instead as an honest broker. Its primary concern has been to prevent an escalation of tensions between two allies that could weaken NATO's cohesion and military effectiveness. These efforts at mediation, however, have often aggravated relations with either Greece or Turkey—and in a number of instances with both.

The 1963–1964 Cyprus crisis provides a good example. President Johnson's letter to Prime Minister Inönü—in which he warned that the United States and NATO might not come to Turkey's aid if a Turkish invasion of Cyprus provoked Soviet intervention—succeeded in preventing a Turkish invasion. But it created a furor in Turkey and prompted Turkey to reduce its dependence on the United States and diversify its foreign policy, including undertaking a major effort to improve relations with Moscow.

Similarly, U.S. sanctions imposed following the Turkish invasion of Cyprus in 1974 led to a sharp deterioration of U.S.-Turkish relations. When the U.S. Congress imposed an arms embargo on Turkey, Turkey responded by temporarily suspending U.S. access to key facilities on Turkish soil. Many Turks regarded the embargo as an unfair slap in the face of a loyal ally and its memory still rankles in many Turkish quarters today. In addition, Greece temporarily withdrew from the military wing of NATO to protest the Turkish intervention and the weak U.S. and NATO reaction to the Turkish invasion.

Although the United States has tried to pursue an even-handed policy and avoid choosing sides, Turkey has always been regarded as the strategically more important ally. During the Cold War, Turkey served as an important bulwark against the expansion of Soviet power into the Mediterranean and the Middle East, tying down some 24 Soviet divisions. It also provided valuable communications and intelligence assets for monitoring Soviet troop movements and verifying arms control agreements.

With the end of the Cold War, Turkey's strategic importance in U.S. eyes has increased.[22] Turkey is at the nexus of three areas of increasing geostrategic importance to the United States: the Caucasus, the Middle East, and the Balkans. In each of these areas, Turkey's cooperation is critical for the achievement of U.S. foreign policy objectives. Moreover, Turkey's strategic weight has increased in U.S. eyes as a result of the war on terrorism. This limits the degree to which the United States is willing to exert pressure on Turkey over issues such as Cyprus and the Aegean.

However, the perception in Greece is quite different. U.S. "even-handedness" is seen in Athens as an example of Washington's willingness to overlook Turkey's violations of international law, especially Turkey's occupation of Cyprus. As Dimitrios Triantaphyllou has noted, "As long as Greeks perceive the United States to be a biased interlocutor between Greece and Turkey and over the Cyprus question, U.S.-Greek relations will continue to be viewed with suspicion in Athens."[23]

However, Washington's ability to influence Turkish policy has significantly declined in the last decade. With the end of the Cold War, Turkey is less in need of U.S. "protection." In addition, Turkey today has foreign policy options—in the Caucasus, Central Asia, the Middle East, and Balkans—that were not open to it a decade ago. It is thus less ready to automatically fall in line behind U.S. policy, especially when U.S. preferences conflict with its own regional interests. The

---

[22]For a fuller discussion, see F. Stephen Larrabee, "U.S. and European Policy Toward Turkey and the Caspian Basin," pp. 143–173.

[23]Dimitrios Triantaphyllou, "Further Turmoil Ahead?" in Dimitris Keridis and Dimitrios Triantaphyllou, *Greek-Turkish Relations in an Era of Globalization*, Dallas, VA: Brassey's, 2001, pp. 73–74.

ending of U.S. military assistance has also served to reduce U.S. leverage over Turkish policy.

At the same time, U.S.-Turkish relations have been increasingly affected by U.S. domestic politics, especially the influence of the Greek-American lobby. Over the last decade, the lobby has been successful in mobilizing support in the U.S Congress to halt or delay arms sales to Turkey on a number of occasions. These delays have been a source of increasing irritation in Ankara and have been one of the reasons behind Turkey's intensified military cooperation with Israel, which Ankara sees as a means of reducing its dependence on American (and European) arms. Ankara has also viewed closer cooperation with Israel as a way of exploiting the political clout of the Israeli lobby in the United States for its own political purposes.

Despite these difficulties, the United States has continued to actively encourage a process of détente and reconciliation between Greece and Turkey. U.S. diplomatic intervention was critical in defusing the crisis over Imia/Kardak in early 1996. The United States also played an important behind-the-scene role in promoting the nonaggression pledge by President Demirel and Prime Minister Simitis at the NATO summit in Madrid in July 1997 and in defusing the crisis over the deployment of the S-300 missiles. More recently, Washington has actively pushed for a dialogue between Athens and Ankara on confidence-building measures.

Cyprus, however, remains an irritant in U.S.-Turkish relations, especially with the U.S. Congress. Turkey's human rights record and continued occupation of Cyprus have prompted the Congress to hold up a number of arms sales to Turkey, causing difficulties in U.S.-Turkish defense relations. In November 2000, for instance, Senator Joseph Biden, the ranking member of the Senate Foreign Relations Committee, temporarily held up export licenses for the sale of eight U.S. CH-53E heavy lift helicopters to Turkey because of Turkey's policy toward Cyprus.[24]

The United States, however, has avoided getting too deeply involved in the Cyprus issue. Instead Washington has encouraged the UN to

---

[24]"Biden Holds Up Export Licenses for CH-53Es for Turkey," *Defense Daily*, November 16, 2000.

take the lead on Cyprus, with the United States playing a low-key "supportive" role behind the scenes. As Morton Abramowitz has noted, "the inclination of every administration is to try to push the Cyprus issue off into the future in the hope that some event—a shift in EU policy, the departure of Denktash, or a change of government in Ankara—will change the context and open up new opportunities for a settlement later."[25]

There have been a few notable exceptions to this pattern. In the summer of 1991, the first Bush administration tried to invigorate the intercommunal talks. But the effort failed to bear fruit and was soon abandoned. After the end of the Bosnian conflict, the Clinton administration seemed about to make a new push for a Cyprus settlement. However, the administration's plans were derailed by the outbreak of the Imia/Kardak crisis and growing domestic instability in Turkey.

In June 1997, Clinton made another attempt, appointing Richard Holbrooke, the architect of the Dayton Accord, as special envoy to Cyprus. Holbrooke made a number of trips to Cyprus in an effort to jump-start the intercommunal talks, but his efforts failed to break the deadlock. The United States played an important behind-the-scene role in getting the "proximity talks" started in December 1999, but since then it has not given the Cyprus issue high priority.

The Bush administration has been too preoccupied with the war on terrorism—and more recently Iraq—to pay much attention to Cyprus. However, this lack of high-level attention is shortsighted. The failure to achieve a Cyprus settlement could lead to a dangerous deterioration of Turkey's relations with the EU and could even stimulate a broader anti-Western backlash among the Turkish population. It could also undermine the recent Greek-Turkish rapprochement.

---

[25]Morton Abramowitz, "The Complexities of American Policymaking on Turkey," in Morton Abramowitz, ed., *Turkey's Transformation and American Policy*, New York: The Century Foundation Press, 2000, p. 164.

THE DOMESTIC DIMENSION

Domestic factors have played an important role in influencing Turkish policy toward Greece and Cyprus. Since the 1990s, Turkey has had a series of weak governments, most of them coalitions. This rapid turnover and lack of a strong government have made the pursuit of bold initiatives toward Greece difficult. With the exception of Turgut Özal, no political leader in Turkey in the last decade has been in a position to make the type of difficult compromises necessary to break the deadlock in relations with Greece. And even Özal's initiatives were ultimately undone by the weakening of his domestic base.

Some important steps toward easing tensions with Greece occurred during Ecevit's second prime ministership. But there was little progress toward a Cyprus settlement. Both Ecevit and Deputy Prime Minister Devlet Bahçeli, the leader of the MHP, opposed any change in Turkish policy on Cyprus, as did the Turkish military. Moreover, the deterioration of Ecevit's health in the spring of 2002 left Turkey leaderless at a critical moment when important decisions needed to be made—especially on Cyprus—and eventually forced Ecevit to call for new elections.

The victory by the Islamist Justice and Development Party in the November 3, 2002, elections adds a new element of uncertainty to Greek-Turkish relations. In the past, rapprochement with Greece has not been high on the AKP agenda—or that of its predecessors—but this may change now that the AKP is in power. The AKP leadership seems to want good relations with Greece. Thus, it is likely to continue the rapprochement with Greece initiated by its predecessors.

Two steps in particular on Turkey's part could help to give Greek-Turkish relations new momentum.

The first would be for Turkey to rescind the parliamentary resolution saying that the extension of Greek territorial waters would be tantamount to a *casus belli*. This resolution has particularly vexed Greek public opinion because Greece has the right under international law to extend its territorial waters to 12 miles but has chosen for political reasons not to do so. A second gesture would be to reopen the theological seminary on the island of Halki which was closed in the early 1970s.

Both moves would give Greek-Turkish détente new momentum and be an important sign of Turkey's commitment to further improving relations with Greece. They would also make it easier domestically for the Greek leadership to justify its détente policy and take additional steps to strengthen it. Indeed, without some reciprocal gestures on Turkey's part, public support for Greek-Turkish détente may be hard to sustain in Greece over the long run.

THE WIDER BALKAN STAGE

The continued differences with Greece over the Aegean and Cyprus have been accompanied by a more active Turkish policy toward the Balkans. Historically, the Balkans have been an area of strong Turkish interest. Turkey is linked to the area by ties of history, culture, and religion. The Balkans were under Ottoman rule for nearly five centuries. This rule left an indelible imprint on the culture, political institutions, and social life of the region.[26] Moreover, many members of the Turkish elite—including Atatürk himself—trace their ancestry back to Ottoman rule in the Balkans.

After the Balkan wars (1912–1913), however, Turkey largely withdrew from the Balkans. Following the founding of the Turkish Republic in 1923, Atatürk discouraged any expression of Pan-Turkism and Turkey carefully refrained from making any irredentist claims over the Turkish and Muslim territories in the Balkans. However, Turkey did participate in the Balkan Pact (1934), which it saw as a hedge against Bulgarian and Italian revisionism.[27] After the end of World War II, Ankara focused its attention on NATO and relations with the West. Although Turkey did make efforts to improve ties with some

---

[26]For an excellent discussion of the Ottoman effect on the Balkans, see Maria Todorova, "The Ottoman Legacy in the Balkans," in L. Carol Brown, ed., *Imperial Legacy. The Ottoman Imprint on the Balkans and the Middle East*, New York: Columbia University Press, 1996, pp. 45–77.

[27]See Mustafa Türkes, "The Balkan Pact and Its Immediate Implications for the Balkan States, 1930–1934," *Middle Eastern Studies*, Vol. 30, No. 1, January 1994, pp. 123–144.

Balkan countries, in general, the region was not high on Turkey's foreign policy agenda.

However, since the end of the Cold War Turkey has "rediscovered" the Balkans. To some extent, this rediscovery has been part of a general broadening of Turkey's foreign policy horizons since the fall of the Berlin Wall.[28] But it has also been influenced by the perceived need to prevent instability in the region from spreading further south and spilling over into Turkey itself. Turkey's policymakers opposed the breakup of Yugoslavia because they feared the implications of secessionism for Kurdish separatism in Turkey and for Turkey's territorial integrity. They oppose Kosovo's independence for the same reason.

Feelings of kinship and a shared history have also been important driving forces behind Turkey's policy, especially toward Bosnia. Many of the Turkish elite trace their origins back to ancestors who fled the Balkans as Ottoman power in the region receded at the end of the 19th century. Moreover, Turkey has given preference in its immigration policies to immigrants from the Balkans. The Turkish elite has tended to view these immigrants as "people like themselves" and felt that Turkey could trust them more easily than other minorities, even if they were not ethnically Turkish.[29]

Since 1990–1991, ties with Albania have been strengthened, especially in the military sphere. Under an agreement signed in 1992, Turkey agreed to help modernize the Albanian army as well as help train Albanian military officers. Ankara also helped to rebuild a naval base at Para Limani on the Adriatic coast, to which it will have access.

Relations with Macedonia have also intensified. Turkey was the first country after Bulgaria to recognize the new Macedonian state. Turkey is also helping to modernize Macedonia's armed forces. In

---

[28]The broader dimensions of Turkey's foreign policy "emancipation" are explored in Ian O. Lesser, "Turkey in a Changing Security Environment," *The Journal of International Affairs*, Vol. 54, No. 1, Fall 2000, pp. 183–198. See also F. Stephen Larrabee, "Turkish Foreign and Security Policy: New Dimensions and New Challenges," in Zalmay Khalilzad et al., *The Future of Turkish Western Relations: Toward a Strategic Plan*, pp. 21–51.

[29]See Kemal Kirişçi, "Disaggregating Turkish Citizenship and Immigration Practices," *Middle Eastern Studies*, Vol. 26, No. 3, July 2000, p. 16.

July 1995, the two countries signed a military cooperation agreement providing for the exchange and training of military experts and joint military exercises. In 1998, Turkey also agreed to give Macedonia 20 of its U.S.-made F-5s as part of its effort to assist the Macedonian armed forces.[30]

The most far-reaching improvement, however, has occurred in relations with Bulgaria. During the Cold War, relations between Ankara and Sofia were marked by considerable hostility, in particular because of Bulgaria's mistreatment of the Turkish minority (about 10 percent of the Bulgarian population).[31] Relations deteriorated dramatically in 1989 when Bulgaria forced 300,000 ethnic Turks to emigrate and confiscated their property.

However, relations improved significantly after the collapse of the communist regime in Sofia in November 1989. Since then, the rights and property of the Turkish minority have been restored and more than half of the 300,000 ethnic Turks forced to emigrate in 1989, have returned to Bulgaria. In addition, several agreements on confidence-building measures have been signed, which have helped to reduce threat perceptions and contribute to better mutual understanding. Today, Turkish-Bulgarian relations are the best they have been since the end of World War II.

Turkey's new activism in the Balkans initially aroused concern in Athens. Many Greeks saw Turkey's more active Balkan policy as an attempt by Turkey to establish a "Muslim arc" on Greece's northern border and as part of a larger strategic plan by Turkey to reassert its former hegemonic role in the Balkans.[32] These concerns were reinforced by Turkey's extensive military modernization plans. With the precedent of the 1974 Turkish invasion of Cyprus in mind, some Greeks worried that Turkey might seek to exploit discontent among Greece's Turkish minority and use it as a pretext to launch an attack against Greece and retake Western Thrace.

---

[30]Umit Enginsoy, "Turkey to Give F-5s to Macedonia," *Defense News*, July 13, 1998.

[31]For a comprehensive discussion, see Kemal Karpat, ed., *The Turks of Bulgaria: The History, Culture, and Political Fate of a Minority*, Istanbul: The ISIS Press, 1990.

[32]See Yannis Valinakis, *Greece's Security in the Post–Cold War Era*, Ebenhausen: Stiftung Wissenschaft und Politik, S-394, April 1994.

However, Turkey's policy in the Balkans has actually been quite cautious. Turkey has not sought to play the "Muslim card," either in Greece or elsewhere in the Balkans. Nor has Turkey shown an inclination to take any unilateral military action in the Balkans. On the contrary, all its military actions in the region have been carried out within a multilateral context, either as part of NATO or United Nations operations. At the same time, Turkey sought to use the crisis in the Balkans as an opportunity to demonstrate its value as a strong NATO ally. Ankara participated in both IFOR and SFOR in Bosnia and provided bases and aircraft for Operation Allied Force in Kosovo. It also contributed 700 troops to the Italian-led Operation Alba in Albania.

Turkey has also taken the lead in the establishment of a multinational peacekeeping force in the Balkans (the Southeast European Brigade, or SEEBRIG). SEEBRIG—which is composed of units from Turkey, Greece, Italy, Romania, Bulgaria, Macedonia, and Albania—has its headquarters in Plovdiv, Bulgaria. Although the brigade is still in its infancy, over the long run it could make an important contribution to promoting greater regional trust and stability.

In short, although Turkey has pursued an active policy in the Balkans since the early 1990s, its policy has been very much in line with that of its NATO allies. Kosovo provides a good example. Despite the fact that Kosovo is almost entirely populated by Muslims and also contains a large Turkish minority (30,000–40,000), Turkey has not taken up the cause of the Kosovars or sought to act as their advocate in international fora. This was in large part because Turkey feared that support for the Albanian Kosovars could legitimize Kurdish separatist tendencies within Turkey.

The Balkans, in fact, have become an area of growing cooperation between Greece and Turkey. During the Kosovo crisis, both countries worked together closely to help dampen and prevent the spread of the conflict. Indeed, both countries have come increasingly to recognize that they share many common interests in the Balkans. This growing convergence of interests has helped to temper the earlier political rivalry in the region and given the recent rapprochement between the two countries greater momentum.

This recent cooperation in the Balkans, however, is highly dependent on a continuation of the process of bilateral détente and progress toward the resolution of other outstanding bilateral issues. If this progress were to be halted, leading to new tensions in bilateral relations, the cooperation in the Balkans between Turkey and Greece could be adversely affected. The emergence of a more nationalistic regime in Ankara could also result in a more assertive, less-cooperative Turkish policy in the Balkans. Finally, a serious deterioration of Turkey's relationship with the EU could diminish Turkey's readiness to cooperate with Greece in the Balkans.

Chapter Five

TURKEY AND EURASIA

Since the early 1990s, Central Asia and the Caucasus have emerged as significant focal points of Turkish policy. This represents an important shift in Turkish foreign policy. Under Atatürk, Turkey consciously eschewed efforts to cultivate contacts with the Turkic and Muslim populations beyond Turkey's borders. In addition, the closed nature of the Soviet political regime and Moscow's sensitivity regarding its control over the non-Russian nationalities made any communication with the peoples of Central Asia and the Caucasus difficult. As a result, after the founding of the Turkish Republic (1923), Turkey had little contact with the peoples of Central Asia and the Caucasus.

The collapse of the Soviet Union, however, created new opportunities—and new challenges—for Turkish policy. With the disintegration of the USSR, a whole new "Turkic world," previously closed to Turkish policy, was opened up.[1] Turkish politicians, especially former President Turgut Özal, saw Central Asia as a new field for expanding Turkish influence and enhancing Turkey's strategic importance to the West. At the same time, the opening to Central Asia and the Caucasus was seen as a way to offset Turkey's difficulties with Europe.

Although Turkey's initial forays into Eurasia met with mixed success—for reasons discussed in greater detail below—the events of

---

[1]For an early assessment of this reawakening, see Graham E. Fuller, "Turkey's New Eastern Orientation," in Graham E. Fuller and Ian O. Lesser, *Turkey's New Geopolitics: From the Balkans to Western China*, pp. 37–97.

September 11 and the U.S.-led war on terrorism are reshaping the geopolitics of Eurasia. Central Asia and the Caucasus have taken on new geostrategic importance, especially for the United States. At the same time, Russia's relations with Turkey have undergone significant change in recent years—a development that is likely to be accelerated by the events of September 11 and the strategic rapprochement between Moscow and Washington.

CENTRAL ASIA

In the first few years after the collapse of the Soviet Union, Turkey embarked on a concerted campaign to expand relations with the newly independent states of Central Asia and tried to become the unofficial leader of the Turkic-speaking states in the region.[2] Ankara opened up cultural centers in the Central Asian republics; it established extensive scholarship programs to allow Central Asian students to study in Turkey; and it expanded its television broadcasts in an effort to extend its cultural influence in Central Asia.

However, Turkey's attempt to expand its influence in Central Asia had only limited success. As a result, Turkey has been forced to scale back many of its grandiose plans. The early euphoria about Central Asia becoming a Turkish sphere of influence has been replaced by a more sober and realistic approach. Central Asia continues to occupy an important place on Turkey's foreign policy agenda, but today there is a greater recognition of the obstacles Turkey faces in trying to expand its influence in the region.

There were several reasons for Turkey's limited success in expanding its influence in Central Asia in the 1990s: First, Turkey lacked the financial means and resources to play a substantial economic and political role in the region. It also overemphasized the economic benefits from its involvement in Central Asia. The countries of the region are poor. Most want economic and financial assistance from Turkey.

---

[2]For a detailed discussion of these early efforts, see Idris Bal, *Turkey's Relationship with the West and the Turkic Republics*, Burlington, VT: Ashgate Publishing Company, 2000. Also see Gareth Winrow, *Turkey in Post-Soviet Central Asia*, London: Royal Institute of International Affairs, 1995; and Philip Robins, "Between Sentiment and Self-Interest: Turkey's Policy Toward Azerbaijan and the Central Asian States," *The Middle East Journal*, Vol. 47, No. 4, Autumn 1993, pp. 593–610.

But few have much to offer in return. As a result, economic cooperation has not expanded as quickly as Ankara had hoped. Indeed, Central Asia is one of the few areas with which Turkey has a trade surplus.

Second, the "Turkish model," with its emphasis on democracy, secularism, and a viable market economy, has found little enthusiasm among the rulers in Central Asia, most of whom are Soviet-era autocrats more interested in maintaining their own personal power than expanding political democracy. In the last decade, the regimes in the region have increasingly gravitated toward greater authoritarian rule rather than greater democracy and political pluralism. The growing threat from radical Islamic groups has reinforced this trend, prompting many of the leaders in the region to introduce more repressive domestic policies.

Third, Turkish officials initially tended to take a rather patronizing approach to relations with the countries of Central Asia, often acting as the "big brother" who knew best. This patronizing attitude did not sit well with many Central Asian officials. Having just emerged from 70 years of Soviet colonialization, the Central Asian elites did not want to replace one form of domination by another. Moreover, Turkish officials often displayed a poor understanding of the social and political realities in the Central Asian countries.

Fourth, Turkey's domestic travails—the Kurdish insurgency, the Islamic challenge, mounting economic problems—limited the amount of attention and resources Turkey could devote to Central Asia. At the same time, Turkey's economic difficulties tarnished the attraction of the Turkish model in the eyes of many Central Asian states. Many Central Asian leaders questioned whether they would truly be better off adopting the Turkish model and opening up their economies.

Finally, Russian influence in the region proved to be more durable than many Turks anticipated. In the initial period after the collapse of the Soviet Union, Russia failed to develop a coherent policy toward Central Asia. This provided Turkey some leeway to make inroads in the region. However, President Putin skillfully exploited the issue of the struggle against international terrorism to strengthen Russia's ties to the states of Central Asia and reassert Russia's influence in the region before the September 11 terrorist attacks on the United States.[3]

Moreover, Russia enjoys certain economic, political, and geographic advantages in Central Asia.

- Many of the regimes in the region are weak, making them easily vulnerable to Russian pressure.

- Many of the countries, especially Kazakhstan, have large Russian minorities. This gives Russia an important political and psychological pressure point.

- The existence of territorial disputes between a number of states in the region, particularly Kazakhstan and Uzbekistan, has enabled Russia to act as a "mediator" and play one Central Asian state off against the other.

- The economies of the region are closely linked to the Russian economy—a legacy of the Soviet era—especially in the energy sector. As a result, the Central Asian states are highly dependent on trade with Moscow.

- Most of the key energy pipelines run through Russia and are controlled by Moscow. Thus, many of the states in the region are dependent on Russia for the transport of their energy resources to the outside world.

- The elites of Central Asia remain highly Russified—a legacy of more than seventy years of Soviet rule. At the summit of "Turkish-speaking states" in Istanbul in April 2001, for example, most of the heads of state spoke Russian not Turkish.[4]

---

[3]For a comprehensive discussion, see F. Stephen Larrabee, "Russia and Its Neighbors: Integration or Disintegration," in Richard L. Kugler and Ellen L. Frost, eds., *The Global Century, Globalization and National Security, Volume II*, Washington, D.C.: National Defense University Press, 2001, pp. 859–874. See also John Dunlop, "Russia Under Putin: Reintegrating the Post-Soviet Space," *Journal of Democracy*, Vol. 11, No. 3, 2000, pp. 39–47.

[4]The two exceptions were President Aliev of Azerbaijan and Turkish President Necdet Sezer. See "Die meisten Staatschefs sprachen russisch," *Frankfurter Allgemeine Zeitung*, April 30, 2001.

Moreover, Turkey has faced some difficulties of its own. Ankara's ties to Tashkent deteriorated after Islamic terrorist attacks on Uzbekistan in February 1999. Uzbek authorities claimed that the terrorists had planned the attacks from Turkish territory and retaliated by closing down a number of Turkish schools and businesses.

THE CHANGING STRATEGIC CONTEXT

However, the events of September 11 have changed the political context in Central Asia and opened new opportunities for Turkish diplomacy, especially in the military field. In recent years, Turkey has stepped up its military assistance to the Central Asian states of the former Soviet Union. Turkey is providing important military assistance to train and equip the Uzbek army in the war against terrorism.[5] This assistance has helped ease the tensions evident in the mid and late 1990s and has given Turkish-Uzbek relations important new impetus.

Turkey has also increased its military assistance to Kazakhstan. In March 2002, Ankara signed a cooperation agreement with Astana providing for collaboration between the Turkish and Kazakh navies and air forces and for training Kazakh cadets in Turkish military colleges.[6] In addition, Turkey has stepped up military assistance to Turkmenistan and Kyrgyzstan.

Ankara has seen the U.S.-led war on terrorism as a means of increasing its political influence in Central Asia and was an early contributor of troops to the Afghan campaign. The Turkish government agreed to send troops to Afghanistan despite widespread opposition among the Turkish public. This decision was dictated by strategic considerations, above all the desire to influence postconflict policy in Central Asia. Turkey's readiness to take over the leadership of the International Security Assistance Force in Afghanistan (ISAF) from Britain—a move strongly supported by the United States—also reflects its desire to play an important role in the region in the future.

---

[5]In 2001, Turkey provided Uzbekistan with military assistance totaling $1.5 million. The assistance planned for 2002 is expected to considerably exceed that figure. See "Turkey equips, trains Uzbek military," *Turkish Daily News*, March 3, 2002.

[6]RFE/RL *Central Asian Report*, Vol. 2, No. 11, March 21, 2002.

Still, Turkey faces important obstacles to playing a broader role in Central Asia. Many of the problems that prevented Ankara from establishing a stronger footprint in the region before September 11 continue to exist—the domestic weakness and illegitimacy of the regimes in the region, the region's economic underdevelopment, widespread graft and corruption; growing popular discontent, the lack of strong civil societies and rule of law, increasing drug trafficking and organized crime, and weak control of national borders. These structural problems, together with Turkey's own financial difficulties, are likely to limit Turkey's ability to significantly expand its influence in Central Asian, despite the close ethnic, religious, and cultural ties that exist with the states of the region.

GROWING STRATEGIC INTEREST IN THE CAUCASUS

In contrast to Central Asia, where Turkey has had only limited success in carving out a larger regional role for itself, Ankara has had much greater success in increasing its influence in the Caucasus (Armenia excepted). Indeed, the Caucasus has emerged as a region of growing strategic importance for Turkey in recent years.

Relations with Azerbaijan in particular have been strengthened. The two countries are linked by strong historical, cultural, and linguistic ties. Azerbaijani intellectuals played an important role in the revival of Turkic national consciousness in the late 19th and 20th centuries.[7] After the Bolsheviks put an end to Azerbaijan's independence, many Azerbaijani leaders fled to Turkey. A further influx occurred after World War II.

After the collapse of the Soviet Union in 1991, Turkey quickly sought to capitalize on the emergence of an independent Azerbaijan. Turkish ambitions were given a big boost by the election of Ebülfez Elchibey as president of Azerbaijan in June 1992. Elchibey was a strong advocate of the "Turkish model" for Azerbaijan. However, Elchibey's ouster in a coup in June 1993—widely believed to have been orchestrated with Moscow's aid—dashed Ankara's initial hopes

---

[7]On the role of Azerbaijani intellectuals in the rise of Turkic consciousness, see in particular Audrey Alstadt, *The Azerbaijani Turks: Power and Identity Under Russian Rule*, Stanford, CA: Hoover Institution Press, 1992. Also see Paul B. Henze, "Turkey and the Caucasus," *Orbis*, Vol. 45, No. 1, Winter 2001, pp. 81–91.

of using Azerbaijan as a springboard for the expansion of its influence in the Caucasus.[8]

However, Elchibey's removal proved to be only a temporary setback. His replacement, Heidar Aliev, a member of the Soviet Politburo in the Brezhnev era, has proven to be a shrewder and more independent-minded leader than many observers expected. Despite Russian pressure, Aliev has increasingly pursued a pro-Western course in recent years and has moved to strengthen ties to Ankara.

The warming of Turkish-Azerbaijani relations has been particularly visible in the military area. Since 1996, Turkey has been actively engaged in the training of Azerbaijan's military officers; it has also helped to modernize the Azerbaijani military education system to bring it in line with NATO standards. An Azerbaijani peacekeeping platoon has been participating in the KFOR as part of the Turkish battalion.

Economic relations have also intensified. In March 2001, Turkey and Azerbaijan signed a set of agreements in which Azerbaijan agreed to supply Turkey with 2 billion cubic meters of natural gas in 2004, increasing to 6.6 billion by 2007 and continuing through 2019. The pipeline to transport the gas will cross Georgia to Erzurum in Eastern Turkey and parallel the Baku-Ceyhan oil pipeline. The pairing off of the two pipelines is intended to increase the commercial profitability of both lines. It also lends added impetus to the construction of the Baku-Ceyhan pipeline and could possibly make Turkey an important transit link in the transport of Azerbaijani gas to Mediterranean and Balkan countries.

Turkey has also sought to strengthen ties to Georgia. In March 1997, Turkey and Georgia signed an agreement on military assistance and cooperation. The agreement envisages the construction of military training centers in Kodori and Gori and a shooting range outside Tbilisi.[9] Turkey has also been helping Georgia with the reconstruc-

---

[8]For a detailed discussion of Turkish policy toward Azerbaijan during this period, see Idris Bal and Cengiz Basak Bal, "Rise and Fall of Elchibey and Turkey's Central Asian Policy," *Dis Politika*, No. 3-4, 1998, pp. 42–56.

[9]*Jamestown Monitor*, Vol. 5, No. 45, March 5, 1999.

tion of the Vaziani military base, and Georgian military personnel have been studying at Turkish military establishments since 1998.

In January 2000, Turkey and Georgia launched a joint initiative to create a "South Caucasus Stability Pact."[10] The pact is designed to increase Turkey's profile in the region as well as enhance Western involvement in the area. By including other Western powers as well as Russia, Turkey, in effect, sought to legitimize Western involvement in the Caucasus as well as to get Russia to view the region as an area of international cooperation rather than its own self-proclaimed sphere of influence. However, the continued dispute between Armenia and Azerbaijan over Nagorno-Karabakh, as well as Moscow's lack of enthusiasm for the proposal, have inhibited the implementation of the plan.

Turkey's relations with Armenia remain strained by the legacy of the massacre of Armenians by the Ottoman forces in 1915–1916.[11] Recent efforts by the Armenian lobby in the United States and in several European countries, particularly France, to introduce genocide resolutions condemning Turkey for the massacre of the Armenians in 1915 have exacerbated these strains and angered the Turkish authorities, who have strenuously disputed the charges. However, recently Turkey has shown a willingness to provide greater access to the Ottoman Archives—a move long urged by scholars—to help defuse the dispute.

Armenia's occupation of Nagorno-Karabakh has not only strained relations with Azerbaijan but also complicated Turkish-Armenian relations. Turkey was one of the first countries to recognize an independent Armenia after the collapse of the Soviet Union. However, the intensification of the Nagorno-Karabakh occupation in the early 1990s resulted in a deterioration of Turkish-Armenian relations. Turkey closed its border with Armenia and suspended efforts to establish diplomatic relations. Since then, Ankara has

---

[10]For an imaginative effort to flesh out the Caucasus Stability Pact, see Sergiu Celac, Michael Emerson, and Nathalie Focci, *A Stability Pact for the Caucasus*, Brussels: Center for European Policy, May 2000.

[11]For a useful historical overview of Turkish-Armenian relations, see Paul B. Henze, *Turkey and Armenia: Past Problems and Future Prospects*, Santa Monica, CA: RAND, 1996.

made a settlement of the Nagorno-Karabakh conflict a prerequisite for a normalization of relations with Yerevan.

To a large extent, Turkey's policy toward Armenia remains hostage to its relations with Azerbaijan. Any easing of Ankara's position on Nagorno-Karabakh would cause strains in relations with Baku. However, the events of September 11 have also had a spillover effect on the Caucasus. With Russia seeking better ties to the United States and NATO, Ankara and Yerevan have begun to quietly explore ways to improve relations.[12] However, a major improvement in relations is likely to occur only after a settlement of the Nagorno-Karabakh dispute.

In addition, the success of Turkey's policy in the Caucasus depends to a large degree on a continuation of the westward-leaning policies pursued by President Aliev in Azerbaijan and President Shevardnadze in Georgia. Both men, however, are in the twilight of their political careers. (Shevardnadze is in his mid-70s and Aliev is nearly 80). Their departure from the political stage could significantly change the political dynamics in the Caucasus and provide new opportunities for Russia to reassert its influence in the region.

THE ENERGY DIMENSION

The emergence of the Caspian basin as a significant source of energy has changed the geopolitics of Eurasia and given Turkish policy toward Central Asia and the Caucasus an important new dimension. Although initial estimates of Caspian oil reserves were highly exaggerated, these reserves are still important and roughly comparable to those in the North Sea.[13] In addition, Kazakhstan, Turkmenistan, and Uzbekistan rank among the world's 20 countries with the largest reserves of natural gas. And if estimates of the gas reserves in Azerbaijan's Shah Deniz fields prove correct, Azerbaijan could emerge as another important source of natural gas in the region.

---

[12]See "Secret talks between Turkey and Armenia," *Turkish Daily News*, June 10, 2002.

[13]Mustafa Aydin, "Turkish Foreign Policy Towards Central Asia and the Caucasus: Continuity and Change," *Private View*, No. 9, Autumn 2000, p. 36. For a comprehensive discussion of the Caspian reserves, see Richard Sokolsky and Tanya Charlick-Paley, *NATO and Caspian Security: A Mission Too Far?* Santa Monica, CA: RAND, 1999, Chapter 6.

The Caspian resources, however, are landlocked. To get the energy to international markets, new pipelines need to be built. How the oil is transported—and through which routes—has important geopolitical implications. As a result, the issue of Caspian energy pipelines has assumed increasing importance. In recent years, a modern version of the 19th century "Great Game" has emerged, with pipelines replacing the railroads as the main means for exerting political influence.[14]

Russia has sought to use the pipeline issue as a means of reasserting its political influence over Central Asia and the Caucasus, insisting that a northern pipeline route from Baku to the Russian port of Novorossiisk on the Black Sea should be the main transit route for the transport of South Caspian oil. This would allow Moscow to exert strategic control over the region's resources and give Russia important political leverage over the policies of the producer countries.

Turkey, backed by the United States, has favored the construction of a pipeline from Baku in Azerbaijan through Georgia to Ceyhan on Turkey's Mediterranean coast. Turkey has pinned its hopes for playing a larger strategic role in Central Asia and the Caucasus on the construction of the Baku-Ceyhan pipeline. Because the route is more expensive than other routes, many analysts and businessmen initially expressed skepticism about Baku-Ceyhan's commercial viability. However, the prospects that Baku-Ceyhan will eventually be built have improved as a result of several developments:

- The engineering study completed in May 2001 has reduced many of the concerns of potential investors about the cost of the project. The new cost estimates—$2.8 billion to $2.9 billion—are higher than the initial $2.4 billion estimate but still within an economically acceptable range.

---

[14]See Ariel Cohen, "The 'New Great Game': Pipeline Politics in Eurasia," *European Studies*, Vol. 3, No. 1, Spring 1996, pp. 2–15. Also M. E. Ahrari, *The New Great Game in Muslim Central Asia*, McNair Paper 47, Washington, D.C.: Institute for National Strategic Studies, January 1996. A number of analysts, however, have argued that the strategic importance of the Caspian region is vastly exaggerated, see Anatole Lieven, "The (Not So) Great Game," *The National Interest*, Winter 1999/2000, pp. 69–80. See also Martha Brill Olcott, "The Caspian's False Promise," *Foreign Policy*, No. 111, Summer 1998, pp. 95–113.

- In March 2001, Kazakhstan's President Nursultan Nazarbaev pledged that oil from Kazakhstan's East Kashagan field would be transported through Baku-Ceyhan. Kazakh oil is considered important because it could make up for any shortfalls in Azerbaijani oil. However, industry spokesmen now claim that reserves in Azerbaijan are sufficient to make Baku-Ceyhan commercially viable even without Kazakh oil.

- The discovery of large gas deposits at the Shah Deniz fields in Azerbaijan in 1999 prompted British Petroleum and the Norwegian company Statoil to change their basic strategy toward Baku-Ceyhan. The prospect of exporting gas to Turkey gave these companies a strong incentive to support Baku-Ceyhan. If the Shah Deniz pipeline runs parallel to Baku-Ceyhan, the costs of Baku-Ceyhan could be reduced.

- The Bush administration has thrown its full support behind the construction of the Baku-Ceyhan pipeline. Moreover, there is little chance that the administration will lift trade sanctions against Iran in the near future. This means that the "Iranian option" favored by many U.S. oil companies will remain effectively closed, giving a big boost to Baku-Ceyhan.

- Russian opposition to the Baku-Ceyhan pipeline has also begun to soften. In May 2002, Russia signed an agreement to transport some of its oil through a pipeline that will connect its main export port, Novorossiisk, with Baku-Ceyhan.[15] This will reduce oil tanker traffic through the Bosphorus, a key Turkish concern.

Construction of the Baku-Ceyhan pipeline began in September 2002. However, Baku-Ceyhan could still face problems. The pipeline construction costs could exceed the projected $2.8 billion to $2.9 billion costs. In addition, Russia could lower tariffs to undercut Baku-Ceyhan's competitiveness. Either move could endanger Baku-Ceyhan's commercial viability and reduce the willingness of investors to support the project.

A third problem would arise if the United States were to shift its policy toward Iran. A thaw in U.S.-Iranian relations would open up new

---

[15]"Russia signs agreement to transport its oil to Ceyhan," *Turkish Daily News*, May 28, 2002.

prospects for shipping Caspian oil via Iran—a route favored by many U.S. oil companies because it is cheaper. This could undercut investor interest in Baku-Ceyhan. Such a shift in U.S. policy, however, seems unlikely in the short term, especially in light of President Bush's characterization of Iran (together with Iraq and North Korea) as part of an "axis of evil."

One critical issue influencing the construction of the pipeline will be Kazakhstan's participation. In March 2001, Kazakhstan's President Nursultan Nazarbaev pledged that Kazakhstan would export oil from the East Kashagan field through Baku-Ceyhan. Since then, however, the Kazakh position has become more ambiguous.[16] Baku-Ceyhan would still be commercially viable without Kazakh oil. But the absence of Kazakh oil would affect Baku-Ceyhan's output and make it difficult to achieve the goal of exporting one million barrels per day peak capacity. Kazakhstan's participation would also make it easier to attract investors.

For Turkey, however, the most pressing problem is not oil but access to new supplies of natural gas. Despite the recent slowdown caused by the Turkish economic crisis, Turkey is the fastest-growing gas market in Europe. According to BOTAS, the Turkish State Pipeline Company, Turkey's demand for natural gas is expected to rise to 53 billion cubic meters per year by 2010 and 82 billion cubic meters by 2020.[17] At the moment, Turkey's use of gas is being constrained by a shortage of supply. Thus, Turkey represents a lucrative market for gas suppliers.

To meet its growing domestic needs, Turkey has signed a number of agreements with potential suppliers. The most controversial of these is the Blue Stream agreement signed with the Russian firm Gazprom in December 1997. Under the agreement, Gazprom will supply Turkey with 16 billion cubic meters of gas per year for 25 years. The agreement is strongly favored by Turkey's Energy Ministry. But it has

---

[16]In a meeting with U.S. Secretary of State Colin Powell in December 2001, Nazarbaev expressed a preference for a pipeline through Iran as the most economical means of transporting Kazakh oil. See Patrick Tyler, "Kazakh Leader Urges Iran Pipeline Route," *New York Times*, December 10, 2001.

[17]Aydin, "Turkish Foreign Policy Towards Central Asia and the Caucasus: Continuity and Change," p. 42.

been sharply criticized in Turkey because of its costs and the fact that it will significantly increase Turkey's dependence on Russian gas.[18]

Critics of Blue Stream also raised questions about the technical feasibility of the project. The Black Sea section of the Blue Stream will be the deepest gas pipeline in the world and will be constructed in one of the most polluted seas in the world. This presents dangers of pipe corrosion from acidity and possible pipe collapse as a result of hydrostatic pressure. The Black Sea is also susceptible to earthquakes, which could not only result in costly repairs but could present environmental hazards.

These considerations led critics to argue that other alternatives should be sought, such as transporting the gas overland from Izobil'noye in Russia through Georgia to Erzurum or by transporting Turkmen gas through the Trans-Caspian Gas Pipeline. However, despite these objections, the Turkish government decided to go through with the Blue Stream deal. The first pipeline was completed in March 2002 and Russia expects to begin delivering gas to Turkey by the end of 2002.

Turkey also signed a gas agreement with Turkmenistan in May 1999 to ship Turkmen gas to Turkey through the Trans-Caspian Gas Pipeline (TCGP). However, there will not likely be enough demand for gas in Turkey to justify and finance both the Blue Stream and TCGP projects. Moreover, the idiosyncratic policies pursued by Turkmenistan's President Saparmurat Niyazov have delayed the initiation of the TCGP and many observers doubt that the pipeline will ever be built.

Energy, especially access to gas, will continue to be a strong driving force behind Turkish policy in Eurasia in the coming decade. Ankara will retain a strong interest in stability in the Caucasus in particular. Any shift in Azerbaijan's or Georgia's policy could significantly affect Turkish interests, especially the future of Baku-Ceyhan and gas supplies from the Shah Deniz fields, and severely undercut Turkish hopes to play a significant political role in Eurasia over the long run.

---

[18]See in particular Ferruh Demirman, "Blue Stream: a project Turkey could do without," *Turkish Daily News*, April 22, 2001.

THE RUSSIAN FACTOR

Russia is likely to be an increasingly important factor in Turkey's policy toward Eurasia in the coming decade. Historically, Russia has been perceived as an adversary and a threat by Turkey. Russian expansionism and championship of Slavic nationalism in the Balkans was the principal cause of the loss of Ottoman territory in the 19th century. Russia and Turkey were also rivals for influence in the Caucasus. The onset of the Cold War reinforced this adversarial relationship. Stalin's efforts to gain control of the Straits after World War II and his claims against the Turkish provinces of Kars and Ardahan prompted Turkey to abandon Atatürk's policy of neutrality and seek membership in NATO.[19]

This historical experience has conditioned Turkish policymakers to regard Russia with considerable suspicion. However, the collapse of the Soviet Union and the emergence of a new post-Soviet space in Central Asia and the Caucasus have had a profound effect on Turkey's relations with Russia.[20] On the one hand, they have sparked new political rivalries as Turkey has sought to expand its influence in Central Asia and the Caucasus—areas where Moscow has strong historical interests. On the other, they have created important new economic interdependencies and prospects for cooperation.

In the last decade, economic cooperation with Russia has expanded significantly. Russia is Turkey's second-largest trading partner and its largest supplier of natural gas. There is also a vibrant "suitcase trade" between Russia and Turkey. Although this trade has declined somewhat since the mid-1990s, it still forms an important part of the unofficial Turkish economy, giving Turkey a strong incentive to keep political relations with Russia on an even keel. Moreover, parts of the

---

[19]For background, see Ferenc A. Vali, *Bridge across the Bosporus: The Foreign Policy of Turkey*, Baltimore, MD: The Johns Hopkins Press, 1971, pp. 172–173.

[20]For a detailed discussion of recent developments in Turkish-Russian relations, see Duygu Bazoglu Sezer, "Turkish-Russian Relations from Adversary to 'Virtual Rapprochement,'" in Makovsky and Sayari, *Turkey's New World: Changing Dynamics in Turkish Foreign Policy*, pp. 92–115. See also her "Turkish-Russian Relations a Decade Later: From Adversary to Managed Competition," *Perceptions*, Vol. VI, No. 1, March–May 2001, pp. 79–98; and "Turkish-Russian Relations: The Challenge of Reconciling Geopolitical Competition and Economic Partnership," *Turkish Studies*, Vol. 1, No. 1, Spring 2000, pp. 59–82.

Turkish business community have developed a strong economic stake in trade with Russia. Turkish construction firms such as GAMA, Tekfen, and ENKA have substantial investments in Russia. They constitute an important domestic lobby for trade with Russia and have been particularly influential in pushing for the construction of the Blue Stream gas pipeline.

The growing Turkish-Russian economic rapprochement is particularly evident in the energy sphere. Russia supplies over 60 percent of Turkey's natural gas. This figure will rise to close to 80 percent with the completion of the Blue Stream pipeline. This growing economic interdependence is beginning to temper traditional Russian attitudes toward Turkey. Increasingly, Turkey is seen more as an important economic partner than as a geopolitical rival. As noted, Moscow has softened its opposition to the construction of the Baku-Ceyhan pipeline. Some Russian companies such as LUKoil and Yukos have even expressed interest in participating in the consortium.

This growing economic rapprochement has been accompanied by changes in the political arena as well. Despite close historic, religious, and cultural ties with Chechnya, Turkey has regarded Chechnya as largely an "internal" Russian problem and has not sought to stoke the fires of Chechen nationalism. Although some Turkish non-governmental groups have sent aid and money to the Chechen rebels, Turkey has cracked down more forcefully on the activities of pro-Chechen militant groups since the takeover of the Swisshotel in Istanbul by pro-Chechen sympathizers in 2001.

One reason for Turkey's restraint on Chechnya is undoubtedly Ankara's concern that Moscow could seek to exploit the Kurdish issue, as it did during the Soviet period. However, Russia has not sought to "play the Kurdish card"—in part out of fear that Turkey could step up support for the Chechen insurgents. Moscow refused, for instance, to provide asylum for PKK terrorist leader Abdullah Öcalan in 1999, a move that contributed to his eventual capture by Turkish intelligence operatives. Indeed, Ankara and Moscow appear to have come to a tacit agreement that they both have much to lose by supporting separatism.

However, Turkey remains wary about Russia's geopolitical ambitions in the Caucasus, especially Moscow's close military ties to Armenia.

In 2000, Russia and Armenia signed a series of defense agreements that broaden defense cooperation and strengthen Moscow's military position in the region. Of particular concern from the Turkish point of view has been Russia's decision to supply Armenia with MIG-29s and S-300 missiles, which will be deployed at Gyumri, one of Russia's two bases in Armenia. However, the rapprochement between Moscow and Washington since September 11 could increase Armenia's room for maneuver and eventually allow Yerevan to reduce its dependence on Moscow.

Russia's policy toward Georgia has also been viewed with concern in Ankara. Moscow has put pressure on Georgia by dragging its feet regarding the withdrawal from its base at Gudauta, which it agreed to vacate by July 1, 2001, at the November 1999 Organization for Security and Cooperation (OSCE) summit in Istanbul. In addition, it has demanded a 14-year period to withdraw from its bases at Batumi and Akhalkalaki; introduced a visa regime for Georgians working in Russia; demanded the creation of a joint police force to patrol areas of the Georgian border with Chechnya; and periodically cut off gas supplies to Georgia.[21]

Russia has also sought to use the separatist tendencies in Abkhazia and South Ossetia to put pressure on Georgia.[22] In June 2002, the Russian Duma amended the law on Russian citizenship to allow residents of Abkhazia and South Ossetia to become Russian citizens—a move viewed by Georgia as tantamount to indirect and disguised annexation.[23] The Duma's action seemed designed to strengthen the rationale for a continued Russian military presence in Abkhazia and keep open the option of possibly detaching the two regions from Georgia at some point.

---

[21]See "Moscou exerce des pressions sur la Géorgie et retarde la fermeture de ses bases militaires," *Le Monde*, August 5, 2001.

[22]For an excellent discussion of the internal and external dynamics of the conflicts in Abkhazia and South Ossetia, see Ghia Nodia, "Turmoil and Stability in the Caucasus: Internal Developments and External Influence," presentation at the conference "Prospects for Regional and Transregional Cooperation and the Resolution of Conflicts," Yerevan, Armenia, September 27–28, 2000, pp. 85–94.

[23]See Vladimir Socor, "The Russian Squeeze on Georgia," *Russia and Eurasia Review,* Vol. 1, No. 2, June 18, 2002, pp. 6–8.

Georgia's independence is critical from the Turkish point of view. A serious shift in Georgia's position back toward Russia could endanger the construction of Baku-Ceyhan. It would also leave Azerbaijan more exposed. As long as President Shevardnadze is in power, there is little likelihood of such a shift. Shevardnadze has pursued an increasingly pro-Western policy in recent years. However, Shevardnadze's term runs out in April 2005 and under the Georgian constitution he can not run again. His departure could lead to renewed internal instability and weaken Georgia's ability to resist Russian pressure.

U.S.-TURKISH STRATEGIC COOPERATION IN EURASIA

Turkey's ability to pursue an active policy in Eurasia will be heavily influenced by the nature and strength of its ties to the United States. Since the mid-1990s, the United States has increasingly emerged as an important player in Eurasia. The United States has given strong diplomatic support to the construction of the Baku-Ceyhan pipeline, as well as the Trans-Caspian Gas Pipeline. Support for an East-West energy corridor has been seen as a way to support Turkish political ambitions while blocking Iran's access to Caspian energy and preventing the reassertion of Russian hegemony in the region.

U.S.-Turkish cooperation has been particularly close in the Caucasus. Washington and Ankara have worked closely to strengthen ties to Georgia and Azerbaijan and encouraged both countries to adopt a stronger pro-Western position. In addition, the United States has lent strong support to Turkey's effort to construct the Baku-Ceyhan pipeline, which it sees as an important means not only to expand Turkey's role in the region but also to strengthen the independence of Georgia and Azerbaijan vis-à-vis Moscow.

The war on terrorism has given this strategic cooperation new momentum. The Caucasus has become an important factor in the struggle against international terrorism. As a result, the United States has stepped up its engagement, especially in Georgia. This increased military engagement, reflected in particular in the dispatch of special forces units to help train Georgian forces to combat Chechen rebels in the Pankisi Gorge along the Georgian-Chechen border, is another example of how the war on terrorism has begun to affect the balance of political power in Eurasia.

In addition, in December 2001, the Bush administration succeeded in getting Section 907 of the Freedom Support Act—which barred direct U.S. government support to Azerbaijan—repealed.[24] The section, introduced under pressure from the American-Armenian lobby, significantly constrained U.S. freedom of action and policy options vis-à-vis Azerbaijan. With the lifting of the ban, U.S. relations with Azerbaijan are likely to receive new impetus.

The removal of the ban could also open up new avenues of cooperation between the United States and Armenia—including in the military area—and allow Armenia to gradually reduce its dependence on Moscow and expand ties to the West. This in turn could create the conditions for a gradual improvement in relations between Armenia and Turkey, although, as noted above, a serious rapprochement is likely to occur only after a settlement of the Nagorno-Karabakh conflict.

IRAN'S ROLE

Iran is potentially an important rival to Turkey for influence in Central Asia and the Caucasus. However, to date Iran has maintained a relatively low profile in Central Asia and the Caucasus and has not made a serious effort to spread its radical brand of Islam. Rather than trying to export revolution, it has concentrated on providing technical and financial assistance and expanding cultural ties.[25]

Iran's policy toward Central Asia and the Caucasus has been mainly driven by geopolitical considerations. Ideological goals such as the promotion of Islam have been of secondary importance.[26] Tehran's primary concern has been to prevent unrest in Central Asia and the Caucasus from spilling over and affecting Iran's own minorities (50 percent of Iran's population is of non-Persian origin and about 30

---

[24]The administration succeeded in getting the ban lifted after the two main Armenian lobbies split on the issue. The mainstream Armenian Assembly of America supported the administration and the Armenian National Committee of America opposed lifting the ban.

[25]See Edmund Herzig, *Iran and the Former Soviet South,* London: Royal Institute of International Affairs, 1995.

[26]See Brenda Shaffer, *Partners in Need. The Strategic Relationship between Russia and Iran,* Washington, D.C.: The Washington Institute for Near East Policy, May 2001.

percent are Azerbaijanis). This is the main reason why Iran supported Armenia against Azerbaijan in the Nagorno-Karabakh dispute. Religious and cultural affinity would have logically suggested that Iran would side with Azerbaijan. However, geostrategic considerations drove Tehran to support Armenia to keep Azerbaijan weak and ensure that Baku would not be in a position to stir up trouble among Iran's large Azeri population.

For the near future, Iran's influence in Central Asia and the Caucasus is likely to be limited for several reasons.

- The secularized Muslim elite in Central Asia and the Caucasus have little sympathy for Iran's brand of radical Islam.

- Iran is likely to be preoccupied with its domestic priorities and with expanding its influence in the Persian Gulf. This will leave it little time and energy to pursue an active policy in Central Asia and the Caucasus.

- Iran must be sensitive to Russian security interests in Central Asia and the Caucasus. Iran's growing reliance on Russian nuclear technology is likely to reinforce this caution.

- Like Turkey, Iran lacks the resources to be a major regional player.

- U.S. efforts to isolate Iran also weaken Iran's ability to play a major regional role, especially in the energy field.

These factors are likely to limit Iran's ability to play a significant role in Central Asia and the Caucasus in the near to medium term. This situation could change, however, if there were a thaw in U.S.-Iranian relations and the United States were to halt—or at least weaken—its efforts to exclude Iran from meaningful participation in the Caspian energy game. In such a case, Iran could become a much more important factor in the Caspian region. However, as noted above, such a reversal of U.S. policy is unlikely in the near future.

Russia's effort to intensify ties to Iran in recent years has been viewed with concern in Ankara.[27] Moscow and Tehran share a common in-

---

[27]On the broader dimensions of this rapprochement, see Shaffer, *Partners in Need: The Strategic Relationship of Russia and Iran.* Also see Eugene Rumer, *Dangerous*

terest in preventing the expansion of U.S. and Turkish influence in the Caucasus and Central Asia. This has been an important impetus for the growing collaboration between the two countries. Both have sought to block the construction of the Baku-Ceyhan pipeline and prevent Ankara and Washington from expanding their influence in the Caspian basin.

Cooperation in the nuclear field has also intensified in recent years. Russia is helping Iran to build a nuclear reactor at Bushehr and in July 2002 Moscow signed a 10-year blueprint for expanding cooperation, which included plans to build five more nuclear reactors in Iran.[28] Although Russia has insisted that its cooperation is limited to civilian development of nuclear energy, the growing nuclear cooperation with Iran has been a source of concern to officials in Washington, Ankara, and Jerusalem and a major irritant in U.S.-Russian relations.

At the same time, there are important obstacles to the development of a broader strategic relationship between Moscow and Tehran. Russia and Iran are potential competitors as oil producers and prospective alternative transit routes for the transport of gas and oil reserves. The two countries are also at odds over the division of the Caspian Sea bed.[29] In addition, closer ties between Moscow and Washington, evident since the September 11 terrorist attacks on the United States, could constrain close collaboration with Iran, particularly in the nuclear field.[30]

Drift: Russia's Middle East Policy, Washington, D.C.: The Washington Institute for Near East Policy, October 2000, Chapter 6.

[28]See Peter Baker, "Russians Assure U.S. on Iran," The Washington Post, August 3, 2002.

[29]Russia, Kazakhstan, and Azerbaijan want the seabed to be divided into national sectors, which would give Iran only 13 percent of the seabed, whereas Iran wants the seabed to be equally divided among the five littoral states, which would give Iran control over 20 percent of the seabed. Turkmenistan's position is less clear but Ashgabat appears to lean toward the Iranian position.

[30]Although Russia insists its cooperation with Iran is limited to civilian development of nuclear energy, there have been some signs that Moscow might be willing to reconsider its plans to continue to build nuclear reactors in Iran, which U.S. officials fear could be used in a covert program to build nuclear weapons. See Steven Lee Myers, "Russia Says It May Reconsider Nuclear Deal with Iran," New York Times, August 3, 2001.

Iran's tense relations with Azerbaijan also are an obstacle to Tehran's ability to play a larger regional role in the Caucasus. At present there is little likelihood that Iran and Azerbaijan will actually go to war, but this possibility cannot be entirely excluded if the idea of reuniting Azerbaijan with the Azerbaijani part of Iran—an idea advocated by Elchibey during his short-lived tenure as president of Azerbaijan—were to gain greater strength. Such a conflict would put Turkey in a difficult position and could create strains in Turkish-Azerbaijani relations.

Iran's threat to use force to expel a British Petroleum–chartered oil exploration vessel in Azerbaijani waters in July 2001 raises broader questions about Iran's longer-term goals in the Caspian region.[31] Whether the move signals a shift in Iranian policy toward more coercive diplomacy or was simply designed to appease hard-liners in the Iranian leadership is not clear. But as Iran acquires progressively more capable missile and perhaps nuclear capabilities, its policy may become less circumspect and restrained. This could give Tehran's policy in the Caspian—and Turkish-Iranian relations—a new, more assertive dynamic.

THE TURKISH-ISRAELI CONNECTION

One interesting aspect of the Eurasian equation has been the development of Israeli policy in the Caucasus and Central Asia and the role played by the Turkish-Israeli connection.[32] In the last few years, Israel has expanded its role in Eurasia, especially with Azerbaijan. Israel strongly supported Azerbaijan in the Nagorno-Karabakh war and relations have warmed considerably since then. Cooperation in the intelligence field has intensified and there are some indications that Israel may have supplied arms to Azerbaijan.[33] Israel has also expanded ties to Uzbekistan and Kazakhstan.

---

[31]For background, see "Les luttes d'influence pour le contrôle des resources de la Caspienne," *Le Monde*, August 5–6, 2001.

[32]See Bülent Aras, "Post Cold War Realities: Israel's Strategy in Azerbaijan and Central Asia," *Middle East Policy*, Vol. 5, No. 4, January 1998, pp. 69–70.

[33]Svante Cornell, "Geopolitics and Strategic Alignments in the Caucasus and Central Asia," *Perceptions*, Vol. IV, No. 2, June–August 1999, p. 119.

There are several reasons for the rapprochement between Israel and the states in Central Asia and the Caucasus. One of the most important has been the fear of Islamic radicalism and a desire to contain Iranian expansion. Another has been Israel's image as a strong, economically prosperous secular state. A third has been the Israeli-American connection. Many Central Asian and Caucasian states see improved ties to Israel as a means of indirectly strengthening ties to the United States.[34] Finally, Israel is seen as an important source of economic assistance and investment by many countries in Central Asia.

At the same time, the Israeli connection has been instrumental in bringing Syria closer to Iran and Russia. Indeed, two axes have begun to emerge in the Caucasus and Central Asia: a pro-Western axis composed of the United States, Turkey, Georgia, Azerbaijan, and (increasingly) Israel and an anti-Western axis composed of Russia, Iran, Armenia, and Syria. These alignments highlight the degree to which alignments in Eurasia, particularly the Caucasus, are beginning to spill over into the Middle East, creating new political geometries and blurring hard and fast distinctions between the two regions.

Indeed, as Bülent Aras has argued, it is becoming increasingly difficult to separate the Caspian region from the geopolitics of the Middle East.[35] Issues such as energy, the proliferation of weapons of mass destruction, and the resurgence of Islam are breaking down the boundaries between the two regions and reinforcing the argument for considering the regions as a geographic and political whole.

At the same time, the events of September 11 and the war on terrorism are weakening old alignments and creating new ones. President Putin's decision to side with the United States in the war on terrorism could weaken the Russian-Armenian-Iranian-Syrian axis. On one hand, it may increase Armenia's room for maneuver and lead to a gradual reduction of Yerevan's dependence on Moscow. On the

---

[34]The Israeli lobby, for example, has been increasingly supportive of Azerbaijan. This has caused a split with the Armenian lobby, with which the Israeli lobby had previously closely cooperated. See David B. Ottoway and Dan Morgan, "Jewish-Armenian Split Spreads on the Hill—Strategic Issues Put Onetime Allies at Odds," *The Washington Post*, February 9, 1999.

[35]Bülent Aras, "The Caspian Region and Middle East Security," *The Mediterranean Quarterly*, Vol. 13, No. 1, Winter 2002, pp. 86–108.

other hand, it could weaken Russian-Iranian cooperation, especially in the nuclear field. Both developments would work to Turkey's advantage.

BLACK SEA ECONOMIC COOPERATION

As part of its growing interest in Eurasia, Turkey has played an active role in promoting closer cooperation in the Black Sea region. This has been reflected in particular in the high priority given to the Black Sea Economic Cooperation (BSEC). Launched by the late Turkish President Turgut Özal in 1989, BSEC is designed to promote private sector activity and stimulate the free movement of goods and services among member states.[36] It also represented a hedge against Turkey's difficulties with the EU at the time.

In June 1992, BSEC's 11 member states[37] formally signed an agreement in Istanbul to promote cooperation in the fields of energy, transportation, communications, information, and ecology. Since then, the group has taken on a stronger institutional identity. In 1994, a Permanent International Secretariat (PERMIS) was established and assumed duties. In June 1998, the group acquired concrete institutional form as the Black Sea Economic Cooperation Organization, giving it a legal basis and allowing it to establish cooperation with other regional and international organizations. In June 1999, the Black Sea Trade and Development Bank was opened in Thessaloniki.

In addition, at Turkey's initiative, a Black Sea Naval Task Force (BLACKSEAFOR), composed of forces from Turkey, Ukraine, Russia, Bulgaria, Romania, and Georgia, has been set up.[38] The force will fo-

---

[36]For a detailed discussion of the origins and evolution of BSEC, see Yannis Valinakis, "The Black Sea Region: Challenges and Opportunities for Europe," *Chaillot Papers 36*, Paris: West European Union Institute for Security Studies, July 1999; also see Ercan Özer, "Concept and Prospects of the Black Sea Economic Cooperation," *Foreign Policy Review*, Vol. XX, No. 1-2, 1996, pp. 75–106; and Nicolae Micu, "Black Sea Economic Cooperation (BSEC) as a Confidence-Building Measure," *Perceptions*, Vol. 1, No. 4, December–February 1996/97, pp. 68–75.

[37]In addition to Turkey, the other members are Albania, Armenia, Azerbaijan, Bulgaria, Georgia, Greece, Moldova, Romania, Russia, and Ukraine.

[38]For background, see Hason Ulusoy, "A New Formation in the Black Sea: BLACKSEAFOR," *Perceptions*, Vol. VI, No. 4, December 2001–February 2002, pp. 97–106.

cus on search and rescue operations, humanitarian operations, anti-mine sweeping, environmental protection operations, and goodwill visits. BLACKSEAFOR may also be available for possible employment in operations mandated by the UN and OSCE.

However, although BSEC has provided a useful forum for discussing regional issues, it has a number of important weaknesses.

- Most of the members are poor and are at very difficult stages of development. This has inhibited effective economic cooperation.

- Geographically, the group is extremely heterogeneous. Some countries, such as Turkey, Russia, and Ukraine, border on the Black Sea; others, such as Armenia, Azerbaijan, and Albania, do not. Thus, the degree of commonality between members is limited.

- Deep-seated antagonisms and ethnic rivalries exist within the group. A number of members have long-standing disputes with their neighbors. These disputes make the development of any serious security component difficult.

- The organization lacks strong procedures for policy coordination. It also lacks strong and effective leadership.

These weaknesses have limited BSEC's usefulness as a mechanism for fostering regional cooperation. As a result, Turkey's interest in BSEC has diminished in recent years. Ankara continues to actively participate in the organization, but the initiative no longer has the same high priority it had during the first half of the 1990s.

DOMESTIC INFLUENCES

Turkey's policy toward Eurasia also illustrates an important broader trend in Turkish foreign policy in recent years: the proliferation of new actors and institutions into the foreign policy arena. This has created a more varied and complex foreign policy environment. Today, foreign policy is no longer solely the prerogative of the Foreign Ministry and military. Domestic factors play an increasingly important role in shaping Turkish foreign policy.

This growing complexity and pluralism have been particularly evident in Turkish policy toward Central Asia and the Caucasus. A variety of ministries and nongovernmental agencies—particularly the Ministries of Culture and Energy—have exerted an important influence on policy toward Eurasia, as have outside interest groups such as the construction industry. Indeed, the Foreign Ministry appears to have largely lost the initiative in determining policy toward Central Asia and the Caucasus.

The Turkish International Cooperation Agency (TIKA), in particular, plays an important role in Central Asia. Set up in 1992, TIKA's prime purpose is to facilitate the activities of businessmen in the Turkic states. TIKA also helps to organize exchanges of youth groups and other groups from Turkey and the Turkic states. The Ministry of Culture and the Office of Religious Affairs have also been active in the Turkic-speaking states in Central Asia, as have agencies such as the Atatürk Language, History and High Culture Council.

In addition, ethnic lobbies have begun to exert a growing influence on policy. There are an estimated five million Turkish citizens of North Caucasian background in Turkey. These groups have collected money and even sent volunteers to fight in Chechnya. Their activities are not controlled by the Turkish government, but they have an important effect on Turkish policy. In January 1996, for instance, Turkish citizens of North Caucasian origin hijacked a Turkish ferry to publicize the plight of the Chechens. Their action received considerable sympathy among the Turkish public. Such actions have complicated relations with Russia and resulted in demands by Moscow that the Turkish government take stronger action to control the activities of pro-Chechen groups in Turkey.

Officially, Turkey has eschewed any effort to promote Pan-Turkism. However, a number of nongovernmental groups advocate a closer association or cultural union encompassing the Turkic states of Central Asia and the Caucasus. The late Alparslan Turkes, the former head of the MHP, actively promoted a Pan-Turkic agenda.[39] Turkes

---

[39]Turkes played a particularly important role in Turkish policy toward Azerbaijan in the period leading up to Elchibey's election in June 1992 and was often used by the Turkish government as an unofficial emissary to the Turkic Republics in Central Asia.

organized annual meetings of Pan-Turkic groups to which representatives of the Soviet Republics were invited. Although these meetings were unofficial, they were often attended by high-level Turkish politicians.

Since Turkes's death in 1997, the MHP has toned down its Pan-Turkism. However, the party continues to emphasize the need to strengthen Turkey's ties to the Turkic states of Central Asia. Many observers worried that the entry of the MHP into the government after the April 1999 elections might lead to a greater emphasis on Pan-Turkic ideas. However, the party has had surprisingly little effect on Turkish policy toward the Turkic states in Central Asia, which continues to be characterized by pragmatism and realism rather than the promotion of Pan-Turkic goals.

The followers of Fethullah Gülen, the Turkish religious leader from the Nurcu sect, also play an important unofficial role in promoting Turkish interests in Central Asia. Gülen's followers have founded more than 300 schools around the world, the majority of them in the newly independent Turkic states of the former Soviet Union.[40] These schools promote a philosophy based on a synthesis of Turko-Ottoman nationalism rather than Islam. They have played a major role in transmitting Turkish cultural values in these countries. Indeed, their influence may be even greater than that of official Turkish policy.

The impact of domestic factors has been particularly evident in Turkey's Caspian energy policy. In the contest for control of energy policy, the Ministry of Energy and the Foreign Ministry have often been on opposite sides of the policy fence. The Energy Ministry has strongly backed the construction of the Blue Stream gas pipeline, whereas the Foreign Ministry and Turkish military have opposed Blue Stream, arguing that it will increase Turkey's dependence on Russian gas and endanger Turkish security.

For details, see Bal and Bal, "Rise and Fall of Elchibey and Turkey's Central Asian Policy," pp. 44–45.

[40]For a detailed discussion of Gülen's role and influence, see H. Hakan Yavuz, "Towards an Islamic Liberalism: The Nurcu Movement and Fethullah Gülen," *The Middle East Journal*, Vol. 53, No. 4, Autumn 1999, pp. 584–605.

Indeed, one main weakness of Turkey's Eurasian policy has been the lack of overall policy coordination and direction. A large number of ministries and quasi-governmental bodies appear to pursue their own agenda with little overall coordination. There has been no clear-cut policy framework providing overall guidance for policy toward the Caucasus and Central Asia. In the absence of such guidance, Turkey's policies toward the region have been dominated by personal whims and personalized connections.[41] This has often resulted in various ministries and agencies working at cross-purposes and hampered the development of a coherent, overarching policy toward the region.

PROSPECTS FOR THE FUTURE

Turkey has found it harder to capitalize on the opportunities opened up by the collapse of the Soviet Union than it had initially anticipated. However, the events of September 11 have added a new dynamic to the Eurasian equation. In Central Asia and the Caucasus, the deck is being reshuffled, with uncertain consequences for politics in both regions. The war on terrorism could lead to a strengthening of U.S engagement in the region and a corresponding diminution of Russia's influence, opening up new opportunities for Turkish diplomacy in both regions.

Whether Turkey will be able to exploit these new opportunities will depend in large part on Turkey's own domestic evolution, particularly its ability to overcome its internal problems. A weak Turkey wracked by internal instability and preoccupied with domestic problems will have little capacity to pursue a coherent policy toward Eurasia. On the other hand, a Turkey that surmounts its internal difficulties would be in a good position to take advantage of the new opportunities opened up by the war on terrorism.

American policy in Central Asia will also be important. If the war on terrorism leads to deeper U.S. involvement in Central Asia, Turkey could be an important beneficiary. But if the United States washes

---

[41]See Aydin, "Turkish Foreign Policy Towards Central Asia and the Caucasus: Continuity and Change," p. 43.

its hands of the region, either as a result of disinterest or a preoccupation with other priorities—as happened in Pakistan and
Afghanistan after 1989—Turkey may find it difficult to make further
inroads in Central Asia and the Caucasus.

Chapter Six

THE MIDDLE EAST AND THE MEDITERRANEAN

Over the last decade, Turkey has become a more important and assertive regional actor, and much of this new activism has been directed toward the Middle East.[1] Ankara is focused more heavily than ever before on events to the south and east, not as an alternative foreign policy orientation but rather as a response to perceived security challenges. With some exceptions, Turks tend to see the Middle East more as a sphere of risk than as a sphere of opportunity.[2] Leaving aside Turkish policy toward Cyprus and the Aegean, addressed in Chapter Four, Ankara also has some emerging challenges and opportunities in the Mediterranean, including those posed by NATO and EU initiatives.

THE POLICY SETTING

Many aspects of Turkish foreign policy are linked to the country's internal politics. This linkage is quite close and direct in the case of developments in the Middle East, which are seen largely through an internal security lens. The most obvious example of this continues to

---

[1]The new Turkish activism in the Middle East is featured in several recent analyses, including Alan O. Makovsky, "The New Activism in Turkish Foreign Policy," *SAIS Review*, Vol. 19, No. 1, Winter–Spring 1999, pp. 92–113; and Malik Mufti, "Daring and Caution in Turkish Foreign Policy," *The Middle East Journal*, Vol. 52, No. 1, Winter 1998, pp. 32–50. On recent Turkish policy toward the region generally, see Philip Robins, *Turkey and the Middle East*, New York: Council on Foreign Relations Press, 1991; and Henri J. Barkey, ed., *Reluctant Neighbor: Turkey's Role in the Middle East*, Washington, D.C.: U.S. Institute of Peace, 1996.

[2]We are grateful to Alan Makovsky for this formulation.

be the linkage between developments in Northern Iraq, Syria, and Iran and Turkey's Kurdish problem.[3] The waning of the PKK insurgency has yet to weaken this linkage, and Turkey's military and civilian leadership remains focused on the containment of separatism through legal and security means and through economic development programs in the southeast of the country. The behavior of Turkey's Middle Eastern neighbors continues to be seen as a significant factor in this struggle. So, too, are the uncertainties introduced by American strategy toward Iraq. There is also a continuing linkage in the minds of many Turks between the phenomenon of Islamism in Turkish politics and the activities of "fundamentalists" in Iran, Saudi Arabia, and elsewhere across the Middle East.

The growth of a dynamic and internationally oriented private sector has led to more diverse patterns of regional engagement. Turkey's business community tends to be secular (although there is a parallel, if less influential, Islamic business sector) and highly supportive of Turkish integration in European and Western institutions. It is not, by and large, a community that looks South or East, culturally or politically. It is nonetheless among the most active proponents of Turkish economic engagement in Eurasia and the Middle East. It is notable that the opening of the Turkish economy and the expansion of the private sector in the Özal years were accompanied by an important opening to North Africa and the Middle East. Özal enhanced relations with the Gulf states, as well as Libya, Iraq, and Iran, attracting Arab capital to Turkey and encouraging Turkish commercial involvement across the region (Turkey's large-scale construction contracts in Libya and elsewhere date from this period).[4] The activity of the Turkish private sector, especially its powerful holding companies, is now a permanently operating factor in Turkey's Middle Eastern engagement.

Energy security is another factor driving Turkish attention to the Middle East. Access to adequate energy supplies at reasonable prices

---

[3]For a critical discussion of the problem and Turkey's response, see Henri J. Barkey and Graham E. Fuller, *Turkey's Kurdish Question*, Lanham, MD: Rowman and Littlefield, 1998.

[4]Kemal Kirisçi, "Turkey and the Muslim Middle East," in Alan Makovsky and Sabri Sayari, eds., *Turkey's New World*, Washington, D.C.: The Washington Institute for Near East Policy, 2000, p. 40.

is now acknowledged as a key factor in Turkey's ability to sustain high growth rates over the coming years—and high growth is seen as essential if Turkey is to converge with European income levels. Over the last decade, Turkey's energy demand has risen by roughly 10 percent per year. Even with Turkey's economic crisis, energy demand is expected to continue its rise, although perhaps at a more modest pace in the near term. Oil accounts for 65 percent and natural gas for over 20 percent of current consumption. Access to oil, although important, is generally seen as less of a concern for Turkey than access to adequate and predictable gas supplies. Gas is an increasingly popular fuel in Turkey, as elsewhere, and supply arrangements are infrastructure intensive and relatively inflexible. Russia is now Turkey's leading gas supplier, but Central Asian and Middle Eastern supplies are likely to become more important over the coming decade. Turkey has very limited domestic energy supplies—satisfying only 3 percent of current usage in the case of natural gas.[5] As a result, Turkish perceptions of the Middle East increasingly feature references to energy security, Turkey's role in Western access, but also access to meet the country's own growing demands.

Western interest in Turkey as a strategic partner is closely bound up with the question of Turkey's role on the European periphery, looking toward Eurasia and especially the Middle East. This is a perspective that sits uncomfortably with Turkey's sense of European identity and policy aspirations. It is also a perspective that has loomed larger after the end of the Cold War and the shift of attention from geostrategic competition in the center of Europe to challenges elsewhere, including the Levant and the Persian Gulf. European and American perspectives affect the Turkish calculus in the Middle East in complex ways but may no longer be a strongly limiting element in Ankara's policy. In this context, the growth of a new and explicitly strategic relationship with Israel offers an important new geometry in Turkish relations with the Middle East.

Overall, the Middle East is likely to continue as a focus for an external policy that has become more active and independent. This, in turn, will make Turkey a more important but potentially more difficult ally

---

[5]William Hale, "Economic Issues in Turkish Foreign Policy," in Makovsky and Sayari, *Turkey's New World*, p. 26.

for the West as it too explores new approaches to regional security and power projection. Growing Turkish military capability and willingness to contemplate regional intervention in defense of national interests will also make Turkey an increasingly significant regional actor in its own right.

A TRADITION OF AMBIVALENCE

Turkey's relations with its Middle Eastern neighbors have long been characterized by mutual ambivalence. During the Ottoman centuries, Turkey was a Middle Eastern power—at times the preeminent Middle Eastern power—with an empire stretching from Arabia to North Africa. The experience of Ottoman rule has left an enduring legacy across the region and has also been reflected in the foreign policy outlook of the modern Turkish republic.[6] The legacy in both cases is an uncomfortable one.

Arab nationalism emerged in large part from the struggle against Ottoman rule, a reality that has left its mark on the outlook of secular nationalists across the region. Arab opinion, especially in Egypt and to a lesser extent elsewhere, tends to regard Turkey as a former colonial power whose regional aspirations should be treated with suspicion. At the same time, Islamists around the Middle East tend to reject the Western orientation of modern Turkey and are understandably hostile to the strongly secular character of the Atatürkist tradition. Some Arab modernizers, notably Bourghiba in Tunisia, have found the Turkish model attractive. But, in general, Turkey and Turkish regional policy have been regarded with suspicion. This tradition has been reinforced by Ankara's membership in NATO and its Cold War alignment with Washington at a time when Turkey's Arab neighbors were either nonaligned or aligned with the Warsaw Pact. The legacy of this modern history can still be felt across a region in which the Cold War generation of intellectuals and leaderships remains largely in place.

---

[6]For a discussion of Turkish policy in the Middle East before and after the formation of the Republic, see William Hale, *Turkish Foreign Policy 1774–2000*, London: Frank Cass, 2000. For an earlier survey, see Vali, *Bridge Across the Bosporus: The Foreign Policy of Turkey*.

Turkish diplomacy has had periods of greater intimacy with the country's Arab neighbors, including a period following the oil crises of the 1970s. These periods of rapprochement have not, however, translated into closer cooperation on issues of importance to Ankara. A leading example is provided by the Cyprus problem. In 1964, and again following the Turkish intervention in 1974, Arab opinion was uniformly negative, despite Turkey's role as protector of fellow Muslims on Cyprus. In the 1960s, Arab support for Greece actually included the supply of arms to the Greek Cypriot militia.[7] In the years since 1974, not one Arab (or Muslim) state has recognized the "Turkish Republic of Northern Cyprus," and states across the Middle East have consistently supported UN resolutions calling for the withdrawal of Turkish forces from the north of the island.

Turkish ambivalence toward the region also operates at several levels. First, the Atatürkist tradition in foreign and security policy was in large measure a rejection of the conditions of weakness and overextension that characterized the Ottoman empire in its last years. The Ottoman presence in Arabia and the Levant was understood as a source of vulnerability, ultimately incompatible with the construction of a modern, unitary Turkish state. The allied defeat of Turkish forces in the Middle East was the proximate reason for Turkish withdrawal, but the Turkish position in negotiations with the allied powers from 1918 onward made clear that the retention of a position in the Middle East was not a priority for the new nation. Even the question of control over Mosul, with its important oil resources, was not pursued as vigorously as it might have been. The strategic priority at the time for the Turkish leadership was the consolidation of national sovereignty within pre-armistice lines. To the extent that territorial and regional issues played a role, Turkey's position in the Balkans and the Caucasus loomed larger (a preference strengthened by the fact that many members of the Ottoman administrative elite had links to these regions). One notable exception to this was the extension of Turkish control over the province of Hatay in the south.[8]

---

[7]Amikam Nachmani, "Turkey and the Middle East," *BESA Security and Policy Studies*, No. 42, 1999, p. 3.

[8]France formally ceded the *sancak* of Alexandretta (Hatay) to Turkey in 1939. Syria continues to dispute Turkish sovereignty over the province.

Taken together, this early experience formed the basis for an enduring, arm's length approach to the Arab Middle East.

Second, the Western orientation of Republican Turkey gave tangible expression to an existing cultural diffidence toward the Arab world. Atatürk's Western outlook has been explained as an attachment to an ideal rather than a specific geographic orientation. The civilization to which Atatürk and his successors have aspired was centered in the West.[9] The Arab Middle East, by contrast, has symbolized Oriental backwardness for generations of Turkish elites (Persian civilization is viewed as a case apart by many Turks, and secular modernization in Iran during the 1930s closely paralleled developments in Atatürk's Turkey). These images of the Arab world have had an enduring influence on Turkish views. They have persisted alongside extensive Turkish commercial activity in the Middle East and are even shared by Turkey's Islamists. When Erbakan made his well-publicized tour of Muslim states shortly after taking office as prime minister in a Refah-led government, not one Arab country was included on the itinerary.

Turkish diffidence regarding the Middle East, especially the Arab world, also affects Turkish interaction with the West. Although aware of the country's role and interests in the region, any suggestion that Turkey is a Middle Eastern rather than a Western country is still greeted with suspicion. "The researcher who says he is in Turkey because he is interested in Middle Eastern politics is quickly informed that he [or she] is in the wrong place."[10] The security policy and strategic studies approaches developed since 1945 have reinforced this preference, particularly in the United States where Turkey's NATO membership invariably marks the country as "European" in foreign and defense policy circles.

Third, Cold War imperatives focused Turkish attention westward. Like NATO's other southern members, Turkey's strategic planning was oriented toward the country's role in the Central European competition between the Alliance and the Warsaw Pact. Turkey had

---

[9]Andrew Mango, *Atatürk: The Biography of the Founder of Modern Turkey*, New York: Overlook Press, 1999, p. 538.

[10]Fuller, "Turkey's New Eastern Orientation," in Fuller and Lesser, *Turkey's New Geopolitics: From the Balkans to Western China*, p. 51.

security concerns outside this setting, including a number in the Middle East, but these were marginal to planning in Brussels and Ankara. And many Middle Eastern questions, including the problem of security in the Persian Gulf, were derivative of larger questions about Soviet intentions and Alliance policy "out-of-area." Close ties between the Soviet Union and Arab neighbors, including Syria, also argued for a cautious approach to the Middle East.[11] In the 1950s, Ankara adopted a more active approach to security cooperation and Alliance-building, including participation with Iraq and Pakistan in the 1955 Baghdad Pact and later with Britain, the United States, Turkey, Iran, and Pakistan in the Central Treaty Organization (CENTO). There were also overtures to Egypt. But these efforts were firmly embedded in East-West rather than regional realities.[12] The tradition of restrained policy and minimal engagement in the Middle East survived multiple Arab-Israeli wars, the Iranian revolution, and Turkey's own Cyprus intervention which was, geographically at least, a Middle Eastern crisis.

THE GULF WAR AND AFTERMATH

The 1990 Gulf War was a watershed in Turkish foreign and security policy, above all in relation to the Middle East. Turkey's approach to the crisis and its aftermath represented a firm break with the past and continues to shape Ankara's regional perceptions. In the years leading up to the Iraqi invasion of Kuwait, Turkish planners were already considering the implications of Turkey's growing economic relationship with Iraq and, in particular, Baghdad's heavy reliance on Turkish pipelines to the Mediterranean for oil exports. This route had acquired greater significance during the Iran-Iraq war in light of

---

[11]Sabri Sayari, "Turkish Foreign Policy in the Post-Cold War Era: The Challenges of Multi-Regionalism," *The Journal of International Affairs*, Vol. 54, No. 1, Fall 2000, p. 170.

[12]For a discussion of Turkey's place in the early Cold War calculus in the Middle East and elsewhere, see Ekavi Athanassopoulou, *Turkey—Anglo-American Security Interests 1945–1952: The First Enlargement of NATO*, London: Frank Cass, 1999. See also, Bruce R. Kuniholm, *The Origins of the Cold War in the Near East: Great Power Conflict and Diplomacy in Iran, Turkey and Greece*, Princeton, NJ: Princeton University Press, 1980.

constraints on shipping from Iraqi ports in the Northern Gulf.[13] At the same time, Turkey was heavily reliant on pipeline revenues as well as energy supplies from Iraq. Indeed, before 1990, Iraq was Turkey's largest trading partner.

In the event, Ankara adopted a very active stance as part of the Gulf War coalition. Iraqi oil exports through Turkish pipelines were cut off as part of UN sanctions. Turkey deployed (with considerable difficulty) some 100,000 troops on the border with Iraq, and allowed air strikes against Iraq to be conducted from Turkish bases, including Incirlik air base near Adana. Several factors were at play behind this forward-leaning policy. The absence of Cold War conditions meant that Turkey could pursue a more active, pro-Western policy without fear of a Russian response. Moreover, the crisis offered an opportunity for Ankara to demonstrate its strategic importance to Europe and the United States in a post–Cold War environment. President Özal himself saw the crisis as offering a window of opportunity to press Turkey's interest in EU membership and to construct a new "strategic relationship" with Washington.

Turkish decisionmaking during the Gulf crisis also reflected changes in civil-military relations in Ankara. Elsewhere in Southern Europe, military establishments adopted an activist stance, arguing for more substantial contributions to coalition military efforts wherever possible. Political leaderships were generally more cautious, preferring symbolic deployments placing few personnel in harm's way. In Turkey, this situation was reversed. President Özal and elements of the civilian political leadership pressed—successfully—for an active diplomatic and military contribution. The Turkish military establishment, including the Turkish General Staff (TGS), pressed for a more cautious policy, fearing the longer-term consequences for Turkey's regional position. Their approach also reflected some concern about the capability of Turkish forces to wage intensive, mobile warfare against Iraqi forces (in fact, this experience was instrumental in spurring Turkey's subsequent defense modernization program). At base, however, the military's attitude reflected a more traditional and measured approach to questions of intervention and national

---

[13]In a conversation with one of the authors in the late 1980s, a senior Turkish officer underlined Turkey's growing control over the Iraqi oil spigot.

sovereignty. Then chief of the TGS, General Torumtay, resigned over the question of Turkish policy during the crisis. In all likelihood, Özal's ability to champion successfully a more assertive pro-Western policy may have been supported by a parallel debate within the military itself about Gulf policy.

Gulf War developments have left an enduring legacy in Turkish policy. Subsequent events have confirmed Turkish perceptions of the region as a source of risk, but the tendency since 1990, and especially since the mid-1990s, has been toward continued activism coupled with greater independence and attention to sovereignty issues. Many Turks view the Gulf War, in particular the establishment of a Western protectorate and no-fly zone in the Kurdish areas of Northern Iraq, as the catalyst for Turkey's decade of conflict with the PKK in Southeastern Anatolia. In this interpretation, the war spurred Kurdish nationalism and also provided a logistical and political opening for the PKK to operate across porous borders with Iraq, Iran, and Syria.

Incredible as this may seem in a NATO context, even some sophisticated Turkish observers will argue that Turkey's allies have deliberately facilitated Kurdish aspirations to foster the breakup of the Turkish state. The suspicions of Western policy regarding Turkey and the Middle East are deeply rooted (Turkish analysts often refer to this as part of the "Sèvres syndrome"—a reference to punitive post–World War I terms that would have imposed draconian territorial and sovereignty concessions on Turkey). The elimination of the Allied ground component in Northern Iraq and the transition from Operation Provide Comfort to the air-only Operation Northern Watch eased some Turkish concerns. But the U.S. and British use of Incirlik air base for the conduct of the operation remains controversial among the Turkish public and politicians.

Turks are fond of saying that the Gulf War had two losers, Iraq and Turkey. By any measure, and despite a good deal of cross-border smuggling, Turkey has lost billions of dollars in pipeline fees and trade revenue from the Iraqi sanctions regime, for which Ankara has never received adequate compensation.[14] Moreover, Turkish policy

---

[14]It has been estimated that the loss of Iraqi trade, including pipeline fees, has cost Ankara some $2 billion per year. Turkey did receive roughly $2.2 billion in compen-

during the Gulf War never produced the immediate benefits in Turkish relations with Europe and the United States that Özal had predicted. The neuralgic issues, from Cyprus to human rights, remained as constraints in relations with the West. Prolonged conflict in Northern Iraq and in Turkey's own Kurdish areas hampered economic development plans and has even become a factor in discussions of the Baku-Ceyhan pipeline.

The experience of the Gulf War clearly reinforced traditional Turkish sensitivities regarding national sovereignty. These sensitivities have been acute in relation to the Kurdish issue and Ankara's conduct of counterinsurgency operations within Turkey and in Northern Iraq. They have also made themselves felt in Turkish policy toward Western intervention in Iraq since 1990. In contrast to the Gulf War, Ankara has been unwilling to allow the use of Turkish bases for offensive air operations against Iraq during any of the subsequent confrontations with Baghdad, including the 1996 crisis over Iraqi operations in the north, and Operation Desert Fox. The rationale for this reserved attitude toward renewed coalition operations is that inconclusive operations against Iraq raise the level of tension with a neighbor with whom Turkey must ultimately coexist. As a matter of public diplomacy, Ankara has been unwilling to participate in the renewed strategic bombardment of Iraq. But Turkey also has serious concerns about the revival of Iraqi military capability, especially WMD and missile programs, and is a quiet beneficiary of the military containment of Iraq. Given fundamental Turkish discomfort with the regime in Baghdad, and the desire for a "seat at the table" in any post-Saddam arrangements for the region, Ankara may feel compelled to support American military intervention in Iraq—but the political and economic price of future cooperation is likely to be high.

CONTOURS OF THE NEW ACTIVISM

In the context of a post–Cold War foreign policy that is generally conservative and multilateral (Ankara's approach to the Balkans is exemplary in this regard), Turkish policy toward the Middle East has

---

satory payments, mainly from Saudi Arabia and Kuwait, in 1991, and a further $900 million in 1992. Cited in Hale, *Turkish Foreign Policy 1774–2000*, p. 225.

been less restrained and more unilateral in character. The contours of this new activism can be seen in recent Turkish policy toward Northern Iraq, Syria, and Israel and to a lesser extent toward Iran.

A FORWARD STRATEGY TOWARD NORTHERN IRAQ

From the mid-1990s, Turkey's strategy toward the Kurdish insurgency has emphasized cross-border operations into Northern Iraq. The aim of these operations has been to deny the PKK sanctuaries in adjacent areas and to ensure that a large proportion of the fighting is carried out on Iraqi rather than Turkish territory.[15] The result has been the establishment of a *de facto* Turkish security zone in Northern Iraq, on the pattern of the Israeli arrangement in Southern Lebanon before the 2000 withdrawal. Together with the increasing proficiency of Turkish forces in counterinsurgency operations, the cross-border strategy contributed to the gradual containment of the PKK threat in its military dimension (it has not made a similar contribution to the development of a satisfactory political strategy to address the Kurdish problem). The frequency of Turkish cross-border operations since 1994 tends to obscure the fact that this kind of intervention would have been almost unthinkable in the pre-Gulf War tradition of Turkish policy toward the region. The scale of these operations, involving as many as 35,000 troops, has also been remarkable.[16] Although the operations have taken place under conditions of murky sovereignty in Northern Iraq, they have changed regional perceptions of the threshold for Turkish action. Indirectly, the Turkish strategy toward Northern Iraq has probably had the effect of strengthening the credibility of Turkish threats to intervene across other Middle Eastern borders in response to internal (i.e., PKK-related) security challenges.

The combination of the battle against the PKK, and the risk of renewed large-scale refugee flows, assures that the situation in North-

---

[15]See Mahmut Bali Aykan, "Turkey's Policy in Northern Iraq 1991–1995," *Middle Eastern Studies*, Vol. 32, No. 4, 1996, pp. 343–366.

[16]Hale, *Turkish Foreign Policy 1774–2000*, p. 309. It was claimed that the cross-border operations in 1995 were the largest military deployments outside Turkey's borders since the foundation of the Republic. The 1974 intervention in Cyprus may cast doubt on this, but the scale remains impressive.

ern Iraq will remain high on the Turkish security agenda. During and after the Gulf War, as many as 1.5 million refugees crossed into Turkey and Iran in response to Saddam's oppression of the Kurds in Northern Iraq. The potential for sudden, renewed refugee movements on Turkey's borders remains a leading concern for Ankara. Indeed, Turkey's support of coalition operations in the Gulf, and later Operations Provide Comfort and Northern Watch, has been motivated at least in part by the desire to monitor and control costly refugee movements.

Strategically, Turkey faces the prospect of continued unpredictability and potential aggression from the regime in Baghdad. American preferences aside, Ankara would probably accept the return of full Iraqi sovereignty in the north as a means of containing residual threats from the PKK and managing the Kurdish problem. Turks have little sympathy for the regime in Baghdad but will continue to find an effective Iraqi government, of whatever stripe, preferable to a political vacuum in Northern Iraq, or worse still in Iraq as a whole, that might foster Kurdish separatism and irredentism. Turkey has a clear economic interest in the reintegration of Iraq and the restoration of the large-scale trade relationship that existed before the Gulf War.

That said, Turkey would clearly prefer to see Iraqi reintegration without a restoration of Iraq's conventional and unconventional military capability. The Turkish military today is in a far better position to address this risk than in 1990 and will be in an even better deterrent position in the coming years. But the reconstitution of Iraqi military power would place Turkey in an uncomfortable position. It would reinforce Ankara's status as a front-line state in the Western confrontation with Baghdad. It would increase the risk of a renewed conflict between Iran and Iraq, a conflict that could destabilize the region and threaten Turkey's economic interests.[17] It would also encourage the European tendency to see Turkey as a barrier to Middle Eastern insecurity, rather than as an integral part of the European security system.

---

[17]For an analysis of Turkish interests and behavior in the Iran-Iraq conflict, see Henri J. Barkey, "The Silent Victor: Turkey's Role in the Gulf War," in Efraim Karsh, ed., *The Iran-Iraq War: Impact and Implications*, London: Macmillan, 1989, pp. 133–153.

Taken together, Turkish preferences with regard to the future of Iraq are not very much different than those of Ankara's American and European allies. But proximity and the close link to Turkish internal security concerns mean that Turkey has a relatively strong interest in regional stability and less interest in the risks inherent in regime change. Again, none of this may stand in the way of Turkish support for a "serious" American effort to change the regime in Iraq.

Iraqi WMD programs pose a special dilemma for Ankara. Turks viewed the Iraqi use of SCUD short-range missiles against Israeli and Saudi targets with alarm and have reacted with concern to periodic Iraqi threats to launch missile attacks on Turkish territory in retaliation for U.S. and British strikes launched from Incirlik as part of Operation Northern Watch. With the progressive extension of ballistic missile ranges across the Middle East, Turkish population centers are now fully exposed to such attacks. For the Turkish military, with its tradition of staunch territorial defense, the inability to counter or deter such threats to the Turkish homeland is particularly worrying. It can be argued that Turkey was exposed to far more extensive missile and WMD risks from the Soviet Union during the Cold War. But that vulnerability was shared with other members of the Alliance and Turks had little reason to doubt the solidity of NATO security guarantees, including the threat of nuclear response. The deterrent situation vis-à-vis Iraq (as well as Iran and Syria) is far murkier, especially in light of NATO's evolving strategy. The emergence of a nuclear Iraq (or Iran) would place all of these issues in very sharp relief. A nuclear breakout on Turkey's borders would pose a range of strategic dilemmas for the West.[18] Even short of new nuclear risks, Turkey will have strong incentives to augment its interest in missile defense with the acquisition of greater deterrent capabilities of its own, including the development of a national missile capability. This, in turn, could affect military balances and strategic perceptions elsewhere, including the Balkans, the Aegean, and the Caucasus.

---

[18]See Michael Eisenstadt, "Preparing for a Nuclear Breakout in the Middle East," *Policywatch* No. 550 parts I and II, Washington, D.C.: The Washington Institute for Near East Policy, August 2001.

A Strategic Relationship with Israel

The development of an overt strategic relationship with Israel offers a different example of Turkish activism in the region. Ankara has had a long-standing, low-key, and cooperative relationship with Israel since the establishment of the Jewish state. From the early 1990s, the relationship began to acquire a more overt and substantive character. Developments since the mid-1990s have moved the relationship into a far more significant realm. Starting with a military training and cooperation agreement in 1996, Ankara has pursued a multifaceted relationship with Israel, ranging from defense-industrial collaboration and intelligence-sharing to economic development and tourism.[19] Plans are also in place for Turkey to export substantial quantities of Manavgat River water to Israel.[20]

The new relationship was facilitated by a period of progress in the Middle East peace process that eased the potentially significant challenges for public diplomacy, both in Turkey where public opinion remains sensitive to the Palestinian problem and across the Middle East. The Turkish-Israeli relationship serves compelling national interests on both sides. Yet observers of the relationship have noted that Ankara is often more open and sweeping than Israel in its description of bilateral ties.

The Turkish rationale is threefold. First, the dominance of the Kurdish challenge at the top of the security agenda in the mid-1990s, and the steadily increasing Turkish concern about Syria's role in support of PKK operations, led Ankara to consider ways of gaining decisive leverage over Damascus. This rationale may have taken on greater urgency with the potential, as it was then seen, for an Israeli-Syrian military disengagement as part of a comprehensive Middle East settlement. This might have left Syria free to concentrate its forces and planning against the Turkish border.

---

[19]For a good survey of the relationship, see Efraim Inbar, "The Strategic Glue in the Israeli-Turkish Alignment," in Barry Rubin and Kemal Kirişçi, eds., *Turkey in World Politics*, Boulder, CO: Lynne Rienner, 2001, pp. 115–126. See also Meliha Benli Altunisik, "Turkish Policy Toward Israel," in Makovsky and Sayari, eds., *Turkey's New World: Changing Dynamics in Turkish Foreign Policy*, pp. 59–73.

[20]See Paul Williams, "Turkey's H2O Diplomacy in the Middle East," *Security Dialogue*, Vol. 32, No. 1, March 2001, pp. 27–40.

Second, despite Turkey's NATO membership and its participation in the Gulf War coalition, Ankara continues to face periodic difficulties in the transfer of arms and military technology from the United States and Europe. Although Western governments remain committed to the support of a strategic ally, European parliaments and the U.S. Congress have been inclined to measure such transfers against Turkey's human rights performance, further complicated by the war against the PKK and outstanding disputes with Greece. By the mid-1990s, Ankara was actively exploring ways to diversify its military procurement against the background of an ambitious modernization program. Russia was an option for unsophisticated systems, and some Russian equipment was purchased for use in counterinsurgency operations where Western scrutiny was greatest. Israel offered an altogether more extensive opportunity for diversification, technology transfer, and training.[21] The importance of these benefits to Turkey is underscored by the Israeli use of Turkish facilities and airspace for training. This activity is highly significant given the general sensitivity of Turkish officials, and especially the Turkish military, to issues of national sovereignty. Some Western observers have actually complained (somewhat inaccurately) that Israel now enjoys better and more predictable access in Turkey than Ankara's NATO allies.

Third, a close and explicit relationship with Israel was seen in some Turkish circles as a way to reinforce the strategic relationship with Washington. This argument is made in the context of American interest in the development of a regional alliance of pro-Western states. A variant holds that the Turkish-Israeli relationship allows

---

[21]The 1996 Military Training and Cooperation Agreement between Turkey and Israel outlined a range of joint training and information-sharing activities, including Israeli access to Turkish airspace for training purposes. Other bilateral agreements provide for technology transfer, joint research, intelligence-sharing, strategic policy-planning talks, and bilateral and multilateral military exercises (on the pattern of the two "Reliant Mermaid" search and rescue exercises held in cooperation with the U.S. Navy in 1998 and 1999). Jordan has participated as an observer in these exercises.

Bilateral arms transfer and defense-industrial agreements have included Israeli modernization of 54 Turkish F-4s for $650 million, a subsequent deal for the upgrade of 48 F-5s, and co-production of the Israeli Popeye II air-to-ground missile. There have also been discussions regarding Turkish participation in the Arrow antiballistic missile (ABM) program, co-production of Merkava tanks, and upgrades to Turkey's aging M-60 tanks. Altunisik, "Turkish Policy Toward Israel," in Makovsky and Sayari, *Turkey's New World: Changing Dynamics in Turkish Foreign Policy*, p. 67.

Turkey to enlist the Israeli "lobby" in support of Turkish interests, including a more stable arms transfer relationship and other matters. This may well be the least persuasive element of the Turkish calculus, reflecting a Turkish preoccupation with the American system of lobbies and an underestimation of complexities on the Israeli side and among Israel's supporters in Washington.

The Turkish-Israeli relationship also serves some additional and shared security concerns related to the containment of Islamic extremism, counterterrorism, and monitoring and countering the proliferation of ballistic missiles and weapons of mass destruction. Turkish strategists take the regional ballistic missile threat very seriously as Turkish population centers are already within range of systems deployed in Iran, Iraq, and Syria. Together with the United States, Israel is a leading source of missile defense technology, and Turkey, Israel, Jordan, and possibly Egypt are likely partners in any U.S.-led theater ballistic missile defense architecture. Turkey is looking to participate in the Israeli-led Arrow missile defense system, although this will require U.S. approval as American-source technology is involved.[22]

How durable is the Turkish-Israeli relationship in light of continuing crises in Israeli-Palestinian relations? Ankara has traditionally pursued an arm's length approach to Arab-Israeli disputes. But the Palestinian issue, in particular, does have resonance for the Turkish public, which remains highly sympathetic to the Palestinian position.[23] Turkish analysts prefer to describe the Turkish approach as "balanced," and this is a reasonable description of Ankara's policy given the initiatives under way with Israel.[24] A good example of this balanced posture can be seen in former President Demirel's participation in the international commission (the Mitchell Commission)

---

[22]Metehan Demir and John D. Morrocco, "Israel, Turkey Eye Joint Missile Shield," *Aviation Week and Space Technology*, July 16, 2001.

[23]See Bülent Aras, "The Impact of the Palestinian-Israeli Peace Process in Turkish Foreign Policy," *Journal of South Asian and Middle Eastern Studies*, Vol. 20, No. 2, 1997, pp. 49–72; and Mahmut Bali Aykan, "The Palestinian Question in Turkish Foreign Policy from the 1950s to the 1990s," *International Journal of Middle East Studies*, Vol. 25, No. 1, 1993, pp. 91–110.

[24]Huseyin Bagci, "Turkish foreign and security policy in 2000: a retrospective," *Turkish Daily News*, December 25, 2000.

established to investigate the nature and handling of violence in Palestinian areas of the West Bank and Gaza.

Since the start of the Arab-Israeli conflict, Turkey has, with rare exceptions, supported Arab resolutions regarding the issue in the UN.[25] At the same time, Turkey has played a role in the post-Madrid peace process, notably through its service as a mentor in Arms Control and Regional Security (ACRS) talks, part of the multilateral track of the process. These negotiations enjoyed considerable success in establishing a set of regional confidence-building measures related to conventional forces. The negotiations, always subject to the vagaries of relations on the bilateral track, foundered on WMD-related questions. By all accounts, however, the Turkish role in facilitating discussion of conventional confidence-building measures (e.g., information sharing, pre-notification of exercises, etc.), informed by Turkey's own experience with the CFE (Conventional Forces in Europe) talks, was exemplary.[26]

The Turkish-Israeli relationship is not immune to the pressure of public opinion, even outside Islamist circles where opposition has always been pronounced. Given the strong interest of the Turkish security establishment, the business community, and secular nationalists in the continuation of the strategic relationship, little short of a full-scale Israeli offensive against the Palestinian Authority is likely to derail current initiatives. Even in the worst case, the cultural and tourism aspects of the relationship may wane, but the military and defense-industrial aspects are almost certain to continue, although perhaps in a lower-key fashion. The rapid growth in bilateral trade, from roughly $100 million in 1991, the year of full Turkish diplomatic recognition, to as much as $2 billion in 2000 (a figure that

---

[25]This pattern has held even after the establishment of full diplomatic relations with Israel in 1991. Turkey supported 170 of 179 UN General Assembly resolutions on Arab-Israeli peace process matters between 1992 and 2000. Alan Makovsky, "Turkish-Israeli Ties in the Context of Arab-Israeli Tension," *Policywatch*, No. 502, Washington, D.C.: The Washington Institute for Near East Policy, November 10, 2000, p. 2.

[26]One of the authors was able to observe Turkey's skillful mentoring role firsthand in 1994 as a member of the U.S. delegation to the ACRS talks, one of the multilateral tracks of the post-Madrid peace process.

includes defense trade), has also created a strong constituency for the relationship within both countries.[27]

Turkey's increasingly substantive relationship with Israel has provoked strong criticism from Turkey's Arab neighbors as well as Iran. Egypt, with its own well-developed sensitivities to the geopolitical balance, and its own expectations of leadership has been especially critical. Periodic Turkish-Israeli and trilateral Turkish-Israeli-U.S. maritime exercises in the Eastern Mediterranean (Reliant Mermaid) have been viewed negatively by Cairo and others. After some initial interest, Jordan is no longer willing to participate in such exercises as an observer. The deepening crisis in Israeli-Palestinian relations has made the issue of Turkish defense cooperation with Israel even more contentious in the Arab world. Yet, as a practical matter, Arab leaderships seem to have reached a grudging acceptance of the Turkish-Israeli relationship as a new strategic factor, and one unlikely to fade under regional pressure.[28]

Confrontation and Rapprochement with Syria

The relationship with Israel may have facilitated a third key demonstration of Turkey's new regional policy: pressure on Syria. Turkish concerns over Syria are long-standing and encompass a number of flashpoints. Syria continues to claim the Turkish province of Hatay and is engaged in a running disagreement with Ankara over the share of downstream waters from the Tigris and Euphrates rivers. Friction over water has been exacerbated in recent years with the completion of the Atatürk Dam and growing Turkish requirements associated with the GAP development project in Southeastern Anatolia.[29] The risk of Turkish-Syrian conflict over either of these issues is remote, although they contribute to a climate of mistrust. A far more serious issue has been Syrian support for the PKK. The PKK leadership, in-

[27]Inbar, "The Strategic Glue in the Israeli-Turkish Alignment," in Rubin and Kirişçi, *Turkey in World Politics,* pp. 115–116.

[28]See Ekavi Athanassopoulou, *Israeli-Turkish Security Ties: Regional Reactions,* Jerusalem: Harry S. Truman Research Institute for the Advancement of Peace, Hebrew University of Jerusalem, March 2001; and Ofra Bengio and Gencer Ozcan, *Arab Perceptions of Turkey and Its Alignment with Israel,* Tel Aviv: BESA Center for Strategic Studies, 2001.

[29]See Williams, "Turkey's H2O Diplomacy in the Middle East," pp. 27–40.

cluding Abdullah Öcalan, had long been resident in Damascus. The Syrian regime, with an eye on its own Kurdish minority, had also provided material support to the PKK, including the use of training bases in the Syrian-controlled Bekaa Valley in Lebanon.[30] In the Turkish view, Syrian support and the infiltration of PKK fighters over the border was a key factor behind the strength of the insurgency on Turkish territory. Ankara lobbied consistently for U.S. and European pressure on Syria to end its support for PKK "terrorism."

Following years of veiled threats of retaliation, including suggestions that Turkey might strike PKK training camps inside Syria and Lebanon, a full-scale crisis erupted in relations with Damascus in the fall of 1998.[31] Turkish officials openly declared their belief that Syria was waging "an undeclared war" on Turkey and that this would bring a Turkish response. The Turkish military deployed reinforcements to the border with Syria against a background of rising tension. The crisis ended in October 1998 with Syrian agreement (the so-called Adana Agreement)—under the pressure of imminent Turkish intervention—to end its support for the PKK and the expulsion of PKK leader Abdullah Öcalan from Damascus.

In addition to providing the first step in the subsequent capture of Öcalan in Kenya, the Adana Agreement has transformed the relationship with Syria. Although some fundamental frictions remain, including Syrian irredentism, water, and proliferation issues, the direct link to Turkish internal security perceptions has been broken, at least for the moment. Turkish officials remain wary of the durability of Syrian commitments to abjure support for the PKK. But Syrian compliance is closely monitored, and Damascus is unlikely to provoke renewed Turkish pressure in a period of regime consolidation and against a background of greater international sensitivity to the sponsorship of terrorist organizations. Moreover, in the wake of the Israeli withdrawal from Southern Lebanon, Syria is under increasing pressure regarding its own presence in that country. Given Ankara's

---

[30]Ankara had already threatened to strike these PKK bases in Lebanon in the early 1990s, and by the mid-1990s discussion of the risk of a hot-pursuit clash with Syria were commonplace in Turkey and the West.

[31]Some Turkish observers, not only Ambassador Sukru Elekdag, had been arguing for a more explicit and tougher strategy toward Syria for some time before the events of 1998.

relationship to Israel, Turkish leverage over Syria has probably never been greater, and the prospects for renewed Turkish-Syrian tension are limited under current conditions. Should the Israeli-Palestinian confrontation escalate to a regional conflict involving Lebanon and Syria, Ankara is likely to be a passive rather than an active participant—but still an unavoidable factor in Syrian calculations.

Looking ahead, Turkey could be a beneficiary of any economic opening in Syria. The practical aspects of the current Turkish-Syrian rapprochement already include steps to facilitate cross-border trade and transport. Under conditions of increased trade between Syria and the EU, as envisaged within the Euro-Mediterranean partnership process, Turkey would be the key link in overland traffic between Syria and Europe, although the Syrian market itself is likely to remain modest. Much will depend on the prospects for an Israeli-Syrian peace agreement in the coming years. Here, too, Ankara has some specific interests. As early as 1986, Özal proposed a "peace pipeline" making Turkish water available in the service of regional peace and development. As a water surplus state in a water-short region, Turkey has much to contribute but also much at stake. Ankara would be a leading beneficiary from a comprehensive settlement but will want reassurance regarding the use of its resources. In this context, there has always been some concern that the regional parties, and above all the United States, will pressure Turkey to offer water-sharing arrangements that might not be in Turkey's best interest (e.g., without adequate compensation).

Ankara will have some additional concerns regarding Syrian missile and WMD programs, as well as the regional military balance in the event of a Syrian-Israeli disengagement. Of all the regional proliferators threatening Turkish territory, Syria has been of the greatest concern for Turkish planners because the Syrian threat has, in Turkish perception, coupled capabilities with intentions.[32] The end of direct Syrian support for the PKK—if durable—may have fundamentally altered this calculus. But Turkish strategists will continue to be wary of chemical and missile developments in Syria. Finally, Turkey will seek

---

[32]It is noteworthy that open briefings on theater missile defense offered by both U.S. and Turkish officials have featured the hypothetical defense of the Turkish port of Iskenderun (a major oil terminus and a key port for NATO reinforcement of Southern Turkey) against Syrian missile attack.

strong assurances regarding limits on the redeployment of Syrian forces in the event of a Syrian disengagement with Israel. In the absence of such limits, Turkish strategists fear that Damascus would be free to reposition its substantial if increasingly obsolete forces to face Turkey.

Friction and Engagement with Iran

To a lesser extent than in relations with Syria, recent Turkish relations with Iran have also shown a propensity for assertiveness. Here, too, there is an important link to Turkish internal security and development concerns, notably Islamism, the Kurds, and natural gas. Since the Iranian revolution, Turkish secularists have been concerned about the prospect for the export of Iranian radicalism. Iran, for its part, has complained about the presence of Iranian opposition groups in Turkey, including elements of the *Mujahiddin-I Khalq*, and alleged cross-border operations. In the 1990s, some prominent terrorist attacks on secular Turkish journalists, intellectuals, and businessmen were thought to have Iranian connections.

The electoral successes of the Refah Party in the early 1990s fueled a debate in Turkey and elsewhere over the extent of links between Turkey's Islamists and Tehran. Many Turkish secularists allege a close connection, including substantial funding. The Erbakan government clearly had an interest in improved relations with Iran as part of a general attempt to shift Turkey's foreign policy gaze eastward, a shift strongly and successfully opposed by the Turkish military and Turkey's foreign policy establishment. Turkey's Islamists have certainly been more interested in developing a close relationship with Iran than in closer ties to the Arab world—a reflection of the preference prevalent in Turkey's religious and secular circles. Yet, in terms of substantive backing, links to supporters in the Gulf, especially Saudi Arabia, have probably been more important in funding the expansion of Islamic activity in Turkey, including the construction of religious schools. Over the past decade, the bulk of the financial support to the Refah (Welfare) and Fazilet (Virtue) Parties, and Turkey's looser Muslim political movements has almost certainly come from traditionally oriented, religious businessmen inside Turkey.

From the mid-1990s, Ankara and Tehran developed a more cooperative relationship, the centerpiece of which was an agreement to
contain the activities of Kurdish insurgents active on both sides of
the border. These and subsequent agreements over PKK operations
have been tenuous and dependent on the vagaries of PKK deployments. Turkish cross-border operations in Northern Iraq and, more
recently, the expulsion of the PKK from Syria have forced the PKK to
operate from sanctuaries in Iran. This has provoked a tough response from Ankara. As early as 1995, Prime Minister Ciller threatened military strikes against PKK bases in Iran.[33] In July 1999, the
Turkish air force reportedly struck PKK camps inside Iranian territory
(Iran claimed that the strikes hit two Iranian villages). Coming in the
wake of Turkish threats to intervene over Syrian support for the PKK,
the Turkish willingness to threaten and use force against PKK targets
in Iran has been interpreted as further evidence of Ankara's new assertiveness in the Middle East.

Energy supply and investment is an increasingly important facet of
Turkish-Iranian relations. Despite periodic frictions over Kurdish issues and Iranian support for terrorism and radical Islam, energy
trade offers a focal point for cooperation, against a backdrop of
growing Turkish concern over the country's energy supply situation.
The Erbakan government signed an agreement that had already been
negotiated by Erbakan's predecessor Tansu Ciller for the import of
Iranian and Turkmen natural gas via a pipeline from Tabriz to Erzurum. With the opening of the Tabriz-Erzurum line in 2002, Iran has
emerged as one of the leading exporters of gas to Turkey, alongside
Russia.

In general, Ankara favors a policy of political and economic engagement toward Tehran and opposes economic sanctions and containment. In this respect, Turkish policy is far closer to the European
than the American approach. Given the potential for economic cooperation, and the importance of bilateral cooperation in policy toward the Kurdish problem, Ankara has a strong interest in Iranian
openness and political reform. Under conditions of reform, together
with a relaxation of American policy, Turkey would be well posi-

---

[33]Hale, *Turkish Foreign Policy 1774–2000*, p. 314.

tioned to play an even more active role in the diplomatic and economic engagement of Iran.

Over the longer term, however, there are some countervailing considerations that could cloud Turkish-Iranian relations. Turks take Iran seriously as a regional actor, and despite points of common interest, Turkey and Iran are essentially geopolitical competitors in the Middle East and Central Asia, including Afghanistan. Iran's nuclear and ballistic missile ambitions—and the Turkish response—will be a central part of this equation. To date, Iranian WMD programs have been overshadowed in the Turkish calculus by more proximate risks from Iraq and, above all, Syria, where proliferation has been combined with multiple flashpoints for conflict. Nonetheless, Iran arguably poses the most serious long-term proliferation risk for Turkey. A nuclear Iran in possession of missiles capable of reaching all major Turkish cities, while holding the territory of Ankara's NATO allies at risk, would fundamentally alter the geopolitical landscape facing Turkey. The need to monitor and counter this threat is almost certainly an important part of the current Turkish-Israeli intelligence and defense relationship. It is a key motivator for Turkish participation in U.S., NATO, and Israeli missile defense initiatives. Indeed, Turkish strategists are already beginning to discuss the utility of a Turkish deterrent in the form of a national missile capability. Much more remote, but not beyond the bounds of credibility, would be the development of a Turkish nuclear capability—unthinkable under current circumstances, but not inconceivable over the coming decades if the NATO nuclear guarantee is uncertain.

THE WESTERN CONTEXT FOR TURKISH REGIONAL POLICY

Turkish-Western cooperation during the Gulf War had an important effect on Western perceptions of Turkey's role in security terms. The war and subsequent crises have strongly reinforced the notion of Turkey as a pivotal actor and a strategic partner. The focus of this interest has been largely Middle Eastern. Ankara's role in successive Balkan crises has redressed this imbalance to some extent. Yet the demonstration of Turkey's Gulf role (reinforced by recent policy toward Israel, Syria, Iran, and Afghanistan) looms behind and may well complicate the question of where and how Ankara fits in emerging European security arrangements. For Turkey, the long-standing fo-

cus on the country's place in the European system may prove difficult to reconcile with a more active role in the Middle East.

Turks have traditionally portrayed the country's role as that of a
bridge between east and west, north and south, and between the
Muslim world and Europe. Europe, for its part, has been more inclined to see Turkey as a *barrier*—a strategic glacis on the European
periphery, holding Middle Eastern risks at bay.[34] This image of
Turkey as a barrier in security terms is reinforced by recent descrip-
tions of Turkey as the West's new "front line" state. But Turkey's desire for closer integration within EU defense arrangements—something the EU has thus far resisted—and Europe's own concerns
complicate this simple picture.

Even in the wake of the EU extension of candidacy status to Turkey at
the December 1999 Helsinki summit, European governments resist
the idea of allowing Ankara to participate fully in foreign and defense
policy decisionmaking. Ankara's tough stance on this issue—discussed in detail in Chapter Three—attests to the strength of the
Turkish conviction that Europe cannot expect to benefit from
Turkey's geopolitical position, including its role in relation to Middle
Eastern risks, if it is unwilling to give Turkey a full seat at the European table.

Europe, for its part, is inclined to recognize the substantial contribution of Turkish diplomacy and military power to security on the
southern periphery. But it is also concerned about Turkish regional
behavior, as well as Turkey's own internal problems. Since the Gulf
War, when some NATO allies questioned the need to deploy even to-
ken reinforcements to Turkey, Ankara has been concerned about the
problem of "selective solidarity" and the growing conditionality of
long-standing security guarantees. These concerns have had an effect on Turkey's own strategy and planning, especially with regard to
the Middle East, where European security commitments are assumed
to be least predictable and where Turkey is increasingly inclined to

---

[34]The "bridge versus barrier" debate continues to have a central place in Turkish
strategic discourse. See Lesser, *Bridge or Barrier? Turkey and the West After the Cold
War.* For a reassessment at the end of the decade, see Ian O. Lesser, "Beyond Bridge or
Barrier: Turkey's Evolving Security Relations with the West," in Makovsky and Sayari,
Turkey's New World: Changing Dynamics in Turkish Foreign Policy, pp. 203–221.

go its own, more assertive way. If Turkey were to become embroiled in a "hot pursuit" incident with Syria or Iran, some European allies might balk at a NATO response.

Some European observers and officials are even inclined to see Turkey as "part of the problem" in the new security environment. Traditional concerns about the movement and status of Turkish workers have been compounded by the position of Turkey as a transit point for illegal migrants seeking entry into Europe from the Middle East and further afield.[35] Turkey is also a major entrepôt for drug smuggling and a variety of international criminal activities affecting Europe. In policy terms, these new challenges should make Turkey a more essential security partner for Europe. In many circles, however, they have only deepened European reservations about Turkey's place in Europe.

In the Turkish perception, the United States has played a very different role as a promoter of the country's strategic importance, particularly in the Middle East and, more recently, in places such as Afghanistan, further afield. As noted above, Ankara does not share all of Washington's policy objectives in the region. Turks are skeptical of the strategy of containment vis-à-vis Iran and are wary of plans for Iraq that appear to foster Kurdish separatism. Ultimately, however, the concentration of hard security challenges on Turkey's Middle Eastern borders, and the longer-term problem of reassurance with regard to Russia, mean that the United States remains the essential strategic partner for Turkey. Indeed, the Russian factor is very much part of the Turkish regional view. Turks view Russia as a potentially serious threat to Ankara's interests from the Balkans to the Gulf. Turks share the American concern about Russia's role in the spread of missile and WMD technology, as well as conventional arms transfers, to Iran. In the event of a sharp deterioration in Russian relations with the West, Moscow is likely to find more room for competition in its policy toward peripheral areas, including the Middle East, than in Europe. This friction would touch directly on Turkish interests.

---

[35]Roger Cohen, "Illegal Migration Increases Sharply in European Union: Istanbul a Transit Point," *New York Times*, December 25, 2000. In 2000 alone, almost 15,000 people were detained trying to enter Greece from Turkey. *Turkiye*, January 2, 2001.

Turkey's role in energy security could become more central to Turkish-Western, and especially Turkish-U.S., relations. Turkey figures prominently in the American debate with regard to power projection in the Caspian and the Gulf (Baghdad is closer to Southeastern Turkey than it is to the lower Persian Gulf), although the record regarding Turkish-U.S. cooperation in Gulf security since the early 1990s is quite mixed. The use of Incirlik air base has been essential to the maintenance of the no-fly zone in Northern Iraq. But Ankara has been very unwilling to facilitate strikes against Iraq proper since the Gulf War. On Iran, as noted above, the Turkish position parallels that of Europe and stresses economic and political engagement. So despite the fact that Turkey's geographic position makes it a potentially important partner for Gulf security—especially if U.S. strategy is realigned to reduce military presence in the Gulf itself—a good deal more, and more effective, joint discussion and policy planning would be necessary for Ankara to accept such a role.

Turkey is directly affected by the progressive extension of the European security "space" to take account of problems emanating from the south, from the Mediterranean, and from the Middle East. NATO's current strategic concept, as well as the tasks envisioned for emerging EU defense arrangements, reflect this trend. Many of the leading contingencies, and some of the most demanding ones, could be on or near Turkey's Middle Eastern borders. Turkey's heightened role in this context was almost certainly a key factor in the EU's strategic decision to offer Turkey candidacy status.

TURKEY AND THE MEDITERRANEAN

Turkish interests in and policy toward the Mediterranean are of a fundamentally different character than in the Middle East. Ankara's contemporary policy toward the region as a whole is conditioned by three factors: the place of Cyprus and the Aegean in relations with Greece, Europe, and the United States; the role of Russia; and the evolving place of the Mediterranean in Western strategic initiatives.

The changing dynamics in relations with Greece are discussed elsewhere in this book (see Chapter Four). But it is important to note a fundamental asymmetry—one of several—in Turkish-Greek relations. Whereas the Mediterranean and the maritime environment generally are central to the Greek geostrategic outlook, the Turkish strategic tradition is essentially continental rather than maritime.[36] This is not to say that Turkey lacks significant interests in the Aegean and the Eastern Mediterranean. Ottoman Turkey was a Mediterranean naval power for hundreds of years, and the Mediterranean itself was a leading battleground in the 500-year competition between the Ottoman empire and the West. Yet, it can be argued that the strategic tradition of modern Turkey, strongly reinforced by the foreign and security policy inclinations of the Republican state, has looked primarily to the risks and opportunities on Turkey's land borders. Overall, the Turkish tradition is far closer to that of Germany and Russia than to the maritime orientation of Britain, the United States, or even Greece. The big strategic challenges, as seen from Ankara, whether in the Balkans, the Caucasus, Central Asia, or the Middle East, have a strongly continental flavor. With the exception of the situation in the Aegean, Turkey's maritime flank is essentially secure.

Turkish strategists do, of course, think about the country's access to Mediterranean Sea lanes for trade and defense. Hence, the great concern over the issue of Greek armaments on "demilitarized" islands in the Aegean. Similar concerns were evident in the Turkish view that new air bases and surface-to-air missiles on Cyprus threatened Turkish freedom of action in the Eastern Mediterranean. As an increasingly important entrepôt for energy and nonenergy trade, Turkey must also be concerned about the free movement of shipping through key choke points in the Eastern Mediterranean—through the Straits themselves, as well as the Suez Canal and the Aegean approaches. In today's strategic climate, threats to these lines of communication are perhaps more likely to come as a result of environmental accidents or terrorism than through conventional interdiction. Restoration of the former Iraqi oil shipments through Turkish pipelines to the Mediterranean, and the realization of the Baku-Ceyhan pipeline scheme, would underscore this interest, as would the expansion of Turkish water shipments from Mediterranean ports. New gas pipelines in the Eastern Mediterranean will also be meaningful in the context of Turkey's own energy interests. Notable developments here include a stalled Israeli-Egyptian agree-

---

[36]See Ian O. Lesser, F. Stephen Larrabee, Michele Zanini, and Katia Vlachos-Dengler, *Greece's New Geopolitics*, Santa Monica, CA: RAND, 2001.

ment for the shipment of Egyptian gas through a pipeline network that could also bring gas supplies to Turkey and a proposal to bring Algerian gas to Greece and Turkey via a pipeline across the Adriatic.

In its historic competition with Russia, Turkey has been concerned about the extension of Russian sea power and influence beyond the Black Sea to the Balkans and the Mediterranean. This concern was at the heart of 19th century Turkish-Western cooperation over the "Eastern Question." It was similarly a central part of the Turkish stake in the Western Alliance throughout the Cold War. Echoes of this concern about Russian activism in the region persist today. The Russian factor was prominent in Turkish perceptions of the S-300 dispute. Officials in Ankara have also viewed with some alarm the Russian presence in the Republic of Cyprus—perhaps as many as 30,000, mostly visitors, in recent years. These elements are sometimes portrayed as part of a wider problem posed by Greek-Cypriot-Serbian-Russian affinity and cooperation—an "Orthodox Axis" threatening Turkish interests and with a natural center in the Eastern Mediterranean. In the wake of Greek-Turkish rapprochement and changes in Belgrade, this notion is now far less popular than at the height of the Balkan crises in the mid-1990s.

Western initiatives in the Mediterranean are another key element in Turkish policy toward the region. Both NATO and the EU have Mediterranean initiatives under way. These can be useful vehicles for Turkish diplomatic and military engagement at a multilateral level in North Africa and the Middle East.[37] Turkey, as a nonmember Mediterranean state, is a participant in the EU's Euro-Mediterranean Partnership program, often referred to as the "Barcelona Process." This initiative has generally been focused on North Africa and the Western Mediterranean. But like other Mediterranean cooperation initiatives, it has acquired a greater stake in the Eastern Mediterranean in recent years. The Barcelona Process has faced considerable difficulty in its economic, political, and security dimensions, and the Southern Mediterranean partners have been highly critical of the process. The evolution of the EU's relationship with Turkey could be viewed as a critical test of the initiative and the future of Eu-

---

[37]See Ian O. Lesser, Jerrold D. Green, F. Stephen Larrabee, and Michele Zanini, *The Future of NATO's Mediterranean Initiative: Prospects and Next Steps*, Santa Monica, CA: RAND, 2000.

rope's overall Mediterranean strategy. But Turkey's position in the Barcelona Process is now greatly overshadowed in both Ankara and Brussels by Turkey's post-Helsinki status as a candidate for EU membership. Despite some continued opportunities for project funding, it is therefore very unlikely that Barcelona will feature prominently in future Turkish policy toward the Mediterranean or the EU.

NATO's Mediterranean Dialogue is potentially more significant from a Turkish perspective. Like all Mediterranean security dialogues, it has suffered from an unavoidable linkage to the state of the Middle East peace process. Nonetheless, the NATO initiative has moved in directions that bring it closer to Turkish interests. The center of gravity in the initiative has shifted progressively to the Eastern Mediterranean, with Israel, Egypt, and Jordan emerging as the most active interlocutors for the Alliance. Ankara has a natural stake in the evolution of these relationships, and Turkey's existing cooperation activities with Israel and Jordan can provide a basis for other multilateral cooperation. There is also a growing interest in moving beyond dialogue to a more extensive program of cooperative activity in the defense realm, including exercises in search and rescue and civil emergency response.[38] Turkey has substantial capabilities in these areas, and could easily host new Alliance programs at its facilities in the region. Like other South European members of the Alliance, Turkey has a broader stake in promoting Mediterranean initiatives as a means of focusing additional attention to security issues in its own backyard. And as a Muslim country, Turkey has a stake in this and other Mediterranean dialogues aimed at forestalling "civilizational" frictions.

LOOKING AHEAD

This analysis suggests a number of overall observations regarding Turkey's role as a security actor in the Middle East. The first and perhaps the most revealing is that Turkey's foreign and security pol-

---

[38]The application of Turkey's security cooperation with Israel and Jordan to the NATO initiative, as well as plans to hold computer and search and rescue exercises in Turkey in 2001, is stressed by TGS officials. TGS briefing at Turkish War Academy, Istanbul, January 24, 2001.

icy establishment sees the Middle East more as a sphere of risk than an area of opportunity. Turkey's economic engagement in the region may increase, and Ankara's diplomatic involvement may wax and wane, but a wide range of security issues, from conventional threats to borders to WMD and refugee flows, will remain at the center of Turkish policy.[39] Questions of policy toward the region will also remain closely linked to questions of internal security—Islamism and the Kurdish problem—especially for Turkey's military leadership.

Second, Turkey's relations with the Arab world are likely to remain ambivalent—at best. Turkish elites, and even Turkey's Islamists, are reluctant to see the Arab Middle East as a natural partner for Turkey. Europe, and the West, will remain the dominant frame of reference across the political spectrum, even if Turkish relations with the EU and the United States are troubled.

Third, the Middle East will nonetheless continue to be a leading area of activism in Turkey's external policy. Turkish policy toward the region will likely be more assertive, less cautious, and less multilateral in character than elsewhere. Again, this is a product of perceived risk and sensitivity to national interest, rather than a product of affinity. It will include the willingness to contemplate military intervention, especially where there is a perceived link to Turkey's internal security.

Fourth, the strategic relationship with Israel is part of this pattern of regional assertiveness and is proving durable even in the face of crises in the Middle East peace process. In the worst case, continued Israeli-Palestinian conflict could force a return to a lower-key approach but with far greater substance than in previous decades.

Fifth, Turkish military restructuring and modernization plans will make Ankara an increasingly important regional security actor in its own right. This suggests that Turkey's future role, coupled with a more assertive approach to diplomacy and the use of force, will go well beyond its traditional one as a facilitator of Western access and

---

[39]For a somewhat different view emphasizing the parallel growth of Turkey's economic and political engagement in the region, see the excellent discussion in Kemal Kirisçi, "The Future of Turkish Policy Toward the Middle East," in Rubin and Kirisçi, *Turkey in World Politics: An Emerging Multiregional Power*, pp. 93–114.

power projection. Indeed, in the absence of a more concerted U.S. approach to Turkey on strategic planning for contingencies in the Gulf and elsewhere in the region, access to Turkish facilities cannot be taken for granted.

Finally, and over the longer term, these observations suggest that Turkey will be neither a bridge nor a barrier in relation to the Middle East but rather an increasingly capable and independent actor—a more significant and possibly more difficult regional ally.

Chapter Seven

TURKEY AND THE UNITED STATES

The relationship with the United States has been a key aspect of Turkey's foreign and security policy since 1945. Despite fears on both sides that this "strategic relationship" would become less strategic and less important with the end of the Cold War, the relationship has retained its significance for both countries. Indeed, the relationship has arguably acquired even greater significance in the post–Cold War strategic environment—a significance underscored by events since the terrorist attacks of September 11, 2001, and the looming confrontation with Iraq. This sustained importance reflects the unsettled character of regions surrounding Turkey and the primacy of these regions in today's security calculus. It is also a reflection of the changes in Turkish society, the influence these changes have had in the way America sees Turkey, and in Turkey's ability to play a larger regional role. In the broadest sense, Turkey's relationship with the United States is also linked to Turkish perceptions of globalization, a phenomenon closely associated with America's political and economic role.

The bilateral relationship remains heavily focused on security matters, and for good reasons given the character of the environment facing Turkey and the proximity of areas where American security interest are engaged. This is particularly true in relation to places such as Afghanistan, the Caucasus, and Central Asia—areas at the nexus of American counterterrorism and regional security strategies. Nonetheless, the relationship faces pressures for diversification, and there are substantial, relatively underdeveloped opportunities to extend bilateral cooperation on investment, trade, and the nonsecurity or "soft" security aspects of regional policy. Turkey's financial woes make the development of this economic dimension of the relationship more urgent and place new demands on both sides.

For Ankara and Washington, the bilateral relationship is increasingly difficult to assess and conduct in isolation. Europe and European institutions are a critical backdrop. The EU is now a far more important factor in both Turkish and American policymaking, and the triangular relationship among Turkey, Europe, and the United States is in flux at many levels. Turkish-EU relations are now more ambitious but highly uncertain. At the same time, a new debate has emerged about the nature of the transatlantic relationship, with critical implications for Ankara. For many Turks, the evolution of the overall Turkish relationship with Europe will have a great influence on perceptions of the United States and Washington's importance as a counterbalance or even a strategic alternative.

Turkey's relationship with the United States at the start of the 21st century is more important, more complex, and less predictable than in previous decades. This chapter explores the changing contours of Turkish and American interests in the relationship, key areas of convergence and divergence, and prospects for the future.

ORIGINS OF A STRATEGIC RELATIONSHIP

The onset of the Cold War was a transforming development in Turkish-U.S. relations, but the bilateral relationship is, of course, much older. Relations with the United States played only a peripheral part in Ottoman engagement—and conflict—with the West in the 19th century. The U.S. naval presence in the Mediterranean is some two hundred years old.[1] But the origins of this presence were in the Western Mediterranean. By the 1820s, however, contact with the United States had increased substantially with the growth of American diplomatic and commercial involvement in the Eastern Mediterranean associated with the "Turkey trade." In marked contrast to modern patterns of energy trade, much of this commerce in the mid-19th century consisted of American exports of petroleum products to the Ottoman empire. The relationship also had its military dimen-

---

[1] The first American naval visit to Turkey took place in Istanbul in 1800.

sions. Ottoman Turkey was a leading purchaser of surplus arms and ammunition from the American Civil War.[2]

Two factors contributed to an arm's length relationship between Ottoman Turkey and America in the 19th century. First, the leading point of American popular and policy interest in the Eastern Mediterranean was support for Greek independence. The American foreign policy establishment, in particular, shared the Philhellenic inclinations of its counterparts in Britain and elsewhere, and American opinion mirrored Europe's in its criticism of Ottoman "backwardness." Second, unlike Britain and France, 19th century America did not give priority to relations with Turkey as a counterweight to Russian ambitions around the Black Sea and the Eastern Mediterranean. At a time when the "Eastern Question" preoccupied European governments, Washington remained largely aloof. Despite significant commercial interests, and a substantial presence by Protestant missionaries, the American strategic interest in Turkey was limited—a striking contrast with the contemporary situation.

Modernizing intellectuals in late Ottoman and early Republican Turkey looked largely to Europe, and above all to France, for models of reform. This inclination was reinforced by wariness of American federalism as a model for Turkish reform.[3] Turkey's strategic alignment in this period was, first and foremost, with Germany. The general thrust of Turkish interest was continental and European, and despite the rapidly growing economic power of the United States, America was only tangentially engaged in areas of Turkish interest. The limited Turkish attention to the United States was largely negative, at least in the early years of the Republic. The provisions of the Treaty of Sèvres, if implemented, would have had draconian implications for Turkish territory and sovereignty. Sèvres was regarded by Turks as Wilsonian in inspiration, and American notions of national

---

[2]See James A. Field, *America and the Mediterranean World 1776–1882*, Princeton, NJ: Princeton University Press, 1969; and Frank Gervasi, *Thunder Over the Mediterranean*, New York: David McKay, 1975.

[3]See Cengiz Candar, "Some Turkish Perspectives on the United States and American Policy Toward Turkey," in Abramowitz, *Turkey's Transformation and American Policy*, pp. 124–125.

self-determination were seen as encouraging Balkan, Kurdish, and Armenian nationalism at Turkey's expense.[4] The legacy of Sèvres— a phenomenon Turkish analysts often refer to as the "Sèvres syndrome"—continues to fuel Turkish suspicions of American strategy toward Turkey and its region. The lasting effects can be seen in the contemporary Turkish debate about American policy in Northern Iraq and the Kurdish issue.

The experience of two world wars heightened American interest in Turkey, but only within limits. In neither the First nor the Second World War was the Eastern Mediterranean a focus of American military engagement. In the Second World War, the Balkans, the Eastern Mediterranean, and the Middle East, where Turkish neutrality mattered, were principally British spheres of responsibility, and Washington actively resisted British efforts to make Southeastern Europe a center of gravity in the conflict.[5] Only in the latter stages of the war, with deepening concern over Soviet ambitions, did relations with Turkey (and relations with Washington for Ankara) loom larger in the strategic calculus.

Containment of Soviet power quickly became the organizing principle for U.S. involvement in and around Turkey. In a formal sense, the Cold War began in the Eastern Mediterranean with Soviet territorial demands on Ankara and the promulgation of the Truman Doctrine to bolster Greece and Turkey. The first NATO "enlargement" embraced Greece and Turkey.[6] From Ankara's perspective, the immediacy of the Soviet threat made the consolidation of Turkey's Western links and, above all, the strengthening of the strategic alliance with Washington, a leading foreign policy priority.

Throughout the Cold War, Ankara and Washington shared a central interest in the containment of Soviet power and in the maintenance of an effective Atlantic Alliance for this purpose. More broadly, the

---

[4]Ibid., pp. 123–124.

[5]That said, American intelligence services were very active in Turkey throughout the war. This story is told in a very engaging manner in Barry Rubin, *Istanbul Intrigues*, New York: McGraw-Hill, 1989.

[6]See Kuniholm, *The Origins of the Cold War in the Near East: Great Power Conflict and Diplomacy in Iran, Turkey and Greece*; and Athanassopoulou, *Turkey—Anglo-American Security Interests 1945–1952: The First Enlargement of NATO*.

two countries have also shared a similar, if not entirely convergent, approach to international affairs. Turkey's internal and geopolitical positions, and the influence of the Turkish military, have fostered a security-conscious approach to policymaking. Cold War imperatives fostered a parallel, security-oriented approach to foreign relations as seen from Washington. Thus, the dominance of security issues in the bilateral relationship has intellectual and political as well as geostrategic roots.

The two countries also share certain additional characteristics in their strategic cultures. These characteristics include a pronounced sensitivity to questions of national sovereignty (far higher than the modern norm in Europe), a low threshold of tolerance for national insecurity and threats to the "homeland," a high threshold for international intervention—and a willingness to act massively and decisively when this threshold is crossed (e.g., for Turkey in Cyprus in 1974 or, more recently, in Northern Iraq). Foreign policy debates in Ankara and Washington are also characterized by a historic tension between a tradition of nonintervention, even isolation, and demands for more active regional engagement.

Tensions in the bilateral relationship over the past decade have encouraged comparisons with a past golden age of relative stability in Turkish-American relations. In reality, such a golden age characterized only the early years of the Cold War, perhaps until the early 1960s. Certainly from the 1962 Cuban missile crisis, in which Washington traded the withdrawal of nuclear-capable Jupiter missiles based in Turkey in symbolic exchange for the withdrawal of Soviet missiles in Cuba, the relationship has experienced periodic and often severe reverses. The 1964 and 1974 Cyprus crises were clearly low points in the bilateral relationship. In both instances, even the defense ties central to the Cold War relationship were severely affected. Indeed, both countries have used security cooperation as a lever in bilateral relations. During the decades of American security assistance to Turkey, threats to withhold this aid, or the defense equipment associated with it, became a feature of congressional approaches toward Cyprus and the Aegean, as well as human rights issues in Turkey. The end of American security assistance, hailed as a sign of maturity in the bilateral relationship, has reduced this form of leverage, although congressional authorization of commercial arms transfers remains a neuralgic issue.

Turkey exercised its own leverage over questions of base access and
support for American power projection. U.S. access to facilities other
than Incirlik air base has been suspended on occasion, most notably
after the imposition of the U.S. arms embargo in 1974. The Turkish
parliament and public opinion have often seen access to Incirlik as a
lever in bilateral disputes. For Turks, the Defense and Economic Cooperation Agreement (DECA), signed in 1969 and periodically revised, established a *quid pro quo* between access to Turkish defense
facilities and U.S. security assistance, most recently interpreted as
"best efforts" with regard to the transfer of American arms. In reality,
this linkage has never been easy or predictable and has given rise to
considerable resentment on the Turkish side and frustration in
Washington. At the start of the Gulf crisis, former President General
Kenan Evren reportedly advised Özal that "unless you have it written
down, you can't trust the United States." Earlier, in a Cold War setting, Evren also reportedly remarked to a German official that "we
only take U.S. aid because we have to. The U.S. uses aid as an in-
strument of pressure. If we go against their wishes, they start saying
they will cut it off. I sometimes ask them, 'Does the U.S. give aid to
have a strong country on the Southern flank of NATO, or as a tool to
make Turkey do as it wants?'"[7]

The apparent smoothness of relations, especially military-to-military
relations during the Cold War, was also a reflection of the routine
character of much bilateral interaction from the 1950s through the
end of the 1980s. Turks in this period were arguably no less sensitive
to the sovereignty issues surrounding base access and other questions of bilateral and NATO concern. But the controversial issues
were well known and there were few surprises in the day-to-day relationship. The key contingencies were, by and large, NATO contingencies concerning the Soviet Union—in short, high consequence
but low probability cases in which Alliance cohesion would be essential. For decades, there were no day-to-day stresses of the sort
imposed by the American air operations from Incirlik as part of Operation Provide Comfort and its successor, Northern Watch. The periodic stresses that did occur—often quite serious—arose from major
political crises in the relationship.

[7]Quoted in Nicole Pope and Hugh Pope, *Turkey Unveiled: Atatürk and After*, London:
Murray, 1997, p. 242.

During the Cold War, but with even greater conviction since the Gulf War, Turkish policymakers and analysts have observed that Washington really does not have a policy toward Turkey *per se*. Rather, from the Turkish perspective, the American approach to Turkey is a by-product of other more prominent concerns—policy toward Russia, Greece, the Balkans, the Caspian, and the Middle East. Leaving aside the accuracy of this observation, there can be little doubt that Turks perceive their relationship with the United States to be derivative of other American interests. Turks have been especially attentive to any signs of a "Russia first" policy in Washington, whether in relation to CFE negotiations in the 1980s or in the Caspian pipeline debate of the 1990s. Arguably, a focus on Turkey's strategic importance—a focus most Turks wish to encourage—makes it inevitable that Washington will often see policy toward Turkey as largely a function of problems in surrounding areas.[8] It is a dilemma Turkey will find hard to avoid. One way of resolving this dilemma would be for U.S. policymakers to view Turkey as a proxy, a regional power to be promoted with a view to more active Turkish intervention in adjacent regions (along the lines of the "Nixon Doctrine," which had sought to cultivate a series of regional proxies, including Iran). Both Americans and Turks, not to mention Turkey's neighbors, would be very uncomfortable with such an approach.

GULF WAR AND AFTERMATH: AN EXPANDED PARTNERSHIP?

The Gulf War was a watershed in terms of Turkish and American perceptions of the bilateral relationship. The crisis in the Gulf unfolded against a background of post–Cold War uneasiness in Turkey about the country's strategic importance in the eyes of the West, and especially in Washington. Observers of President Özal's policy during the crisis stress that he saw the opportunity for Turkey to play an active role in the Gulf coalition as a chance to reassert Turkey's geopolitical significance in the broadest sense and to reinvigorate the strategic

---

[8]Alan Makovsky, "Marching in Step, Mostly!" *Private View*, Spring 1999, Vol. 3, No. 7, p. 38.

relationship with the United States.[9] Many in Turkey were less en-
thusiastic about Turkish participation in the Gulf War, including senior elements of the Turkish military. Özal and others, however, saw
a chance for Turkey to secure a seat at the table after Baghdad's defeat. At American urging, Ankara granted access and overflight rights
for American combat and supply aircraft operating from Incirlik,
Batman, and elsewhere. Iraqi oil exports through Turkish pipelines
were shut down as part of the economic sanctions against Baghdad.
With considerable difficulty, some 100,000 Turkish troops were
eventually redeployed to the border with Iraq, pinning down substantial Iraqi forces.

The Turkish contribution to the coalition effort in the Gulf was substantial and achieved at some political cost inside Turkey. But far
from the new strategic relationship Özal had envisioned, the Gulf
War and its aftermath have left a legacy of complexity and resentment in bilateral relations. Turks point to the tangible costs of their
support for American aims in the Gulf, including refugee pressures
and a deadly Kurdish insurgency, the loss of revenue from trade with
Iraq, and the sovereignty compromises associated with continued
American (and British) air operations over Northern Iraq conducted
from Incirlik. Objectively, some of these undoubted "costs" to
Turkey might not have been avoided through Turkish neutrality in
the Gulf War. And without U.S. intervention and, ultimately, the security guarantee to Turkey, the costs to Turkey might have been far
higher.

Nonetheless, the Gulf War episode has left many Turks with a sense
of disappointment and suspicion regarding American policy. With
the end of U.S. security assistance, and with economic sanctions
against Iraq still in place more than a decade after the invasion of
Kuwait, many Turks feel that they have little to show for their cooperation with Washington and Europe in the Gulf. Indeed, the primacy of the threat of Kurdish separatism in Ankara's strategic perceptions has meant that U.S. policy toward the Kurdish areas of
Northern Iraq is treated with exceptional suspicion—suspicion that
draws on the deeply rooted "Sèvres syndrome" noted above. In this

---

[9]See, for example, the analysis of former U.S. ambassador in Turkey, Morton
Abramowitz, in Pope and Pope, *Turkey Unveiled: Atatürk and After*. Also, Abramowitz, "The Complexities of American Policymaking on Turkey," pp. 3–35.

climate of suspicion, it is not surprising that each parliamentary extension of the Operation Northern Watch mandate is accompanied by considerable debate and uncertainty.

The Gulf crisis unfolded at a time when many of the traditional underpinnings of the bilateral relationship had disappeared or were under strain. The Cold War context for the American military presence in Turkey had evaporated. Security assistance, with the exception of arms "cascaded" to Turkey by NATO allies under CFE treaty provisions, was in sharp decline. The result was a perceived loss of leverage on both sides and a disinclination to go beyond the rhetoric of enhanced cooperation. Efforts to diversify the bilateral relationship, including a Turkish proposal for a free trade agreement with the United States, were not seriously entertained.

Issues raised during the Gulf War remain sensitive points in the bilateral relationship, with pronounced differences between Washington and Ankara on the diplomatic and economic engagement of Iraq and on the use of force against Baghdad. Ankara has a stake in the containment of Iraqi military power but has been most reluctant to support military strikes against Iraq. In 2001, and over strong American objections, Turkey reestablished full diplomatic relations with Baghdad. At a time when Turkey's financial problems place a premium on backing from Washington, Turkey's stance on Iraq is likely to be a key measure of the health of the bilateral relationship from the perspective of the administration in Washington.

If the Gulf experience has been difficult for both sides, other aspects of the bilateral relationship since 1991 have been more positive. Turkey has been a strong supporter of American policy in the Balkans, both in Bosnia and in Kosovo. Turkey has expressed its willingness to contribute forces to any NATO peacekeeping deployment in Macedonia and favors the presence of American forces alongside those of European allies. Ankara's moderate and multilateral approach to the region has allayed American fears of Turkish friction with Greece over Balkan policy. In the Aegean, where the risks of Greek-Turkish brinkmanship have been a special concern for the United States (it is widely believed that only last minute intervention by Washington prevented a military clash over Imia/Kardak in 1996), the development of a new détente between Ankara and Athens has improved the climate on a key bilateral issue. This improvement has been slower to affect attitudes in the U.S. Congress, where the change in mood has lagged behind changes in the region itself. Ultimately, however, a durable improvement in Greek-Turkish relations is likely to influence the climate surrounding arms transfers and other questions where Aegean balances have been a concern. Movement on Cyprus, less likely, would be a transforming development in this regard and would defuse much of the standing criticism of Turkey in congressional circles.[10]

Developments within Turkey are potentially the most important determinant of how the bilateral relationship will evolve. Here, the decade of the 1990s saw considerable positive change. In the aftermath of the Gulf War, Ankara's counterinsurgency operation in the Southeast and in Northern Iraq kept American attention focused on human rights abuses and the lack of progress on political solutions to the Kurdish problem in Turkey. Support for Turkish democratization and human rights has been a consistent theme of American policy, with successive American officials urging Ankara to "take risks for reform."[11] Behind this approach has been the valid assumption that only a more fully democratic and open Turkey will be in a position to achieve a durable and enhanced strategic partnership with the United States. This is a natural reflection of the post–Cold War realities in American foreign policy. But it sits uncomfortably with the more nationalistic mood in Turkish politics since the Gulf War.

This reflexive nationalism was demonstrated very clearly in the Turkish reaction to the near passage of an Armenian genocide resolution in the U.S. Congress at the end of 2000. This nonbinding resolution was withdrawn at the last moment after intense pressure from the Clinton administration (a similar resolution was adopted by the French parliament a few months later). Armenian genocide resolutions had been introduced periodically in the past, but the impending American elections made this a special case for all sides. Ankara clearly viewed the resolution issue as a key test of the bilateral relationship, with prominent Turkish politicians threatening wide-

---

[10]See Ian O. Lesser, *Turkey, Greece and the U.S. in a Changing Strategic Environment: Testimony Before the House International Relations Committee, Subcommittee on Europe*, Santa Monica, CA: RAND, June 2001.

[11]This theme was articulated with special vigor by Ambassador Marc Grossman in the mid-1990s.

ranging retaliation, including an end to American use of Incirlik air base for non-NATO purposes in the event of passage. The vulnerability of the bilateral relationship to such disturbances, despite its manifestly strategic importance for both sides, is an indication of the delicately poised mood in Turkey and the weight of suspicion just below the surface.

TURKISH BILATERAL INTERESTS TODAY

If the bilateral relationship is often characterized as strategic for the United States, the Turkish stake in the relationship is no less strategic, perhaps even more so. In this context, Ankara has multiple interests.

First, the need for deterrence and reassurance in relations with Russia is deeply imbedded in Turkey's strategic culture and is an element of continuity in the country's geostrategic perceptions and in relations with the West. Concern about Russian intentions, as well as risks flowing from chaos within or around Russia, ranks high on Turkey's security agenda. As a longer-term worry, it is probably at the top of this agenda. In many respects, the Turkish view of Moscow is the most wary and security-oriented in NATO. Turkey no longer shares a border with Moscow, but political and security vacuums in the Caucasus and Central Asia offer new flashpoints for competition and conflict. In the event of a general deterioration in Russian-Western relations, Turkey would be on the front line in a competition far more likely to be focused on the periphery, including Russia's "near abroad," the Balkans and the Middle East—that is, in Turkey's neighborhood—than in the center of Europe. Here, Ankara views the expansion of Russian military sales to Iran and elsewhere in the Middle East with alarm. Ankara also sees Washington as having the primary responsibility and ability to constrain Moscow's arms and technology transfers to the region.

The strategic relationship with the United States and the NATO security guarantee (the two have traditionally been closely linked in Turkish perception) remain indispensable in relation to Russian risks. The NATO nuclear guarantee is still an essential part of this equation for Ankara. Turkish attitudes toward nuclear questions are among the most conservative in NATO, because these questions are seen against a backdrop of heightened concern about Russia and

WMD and ballistic missile risks emanating from the Middle East. In each case, the United States, together with Israel, is Turkey's key partner in the management of these problems.

Ankara seeks a more active role in emerging European defense arrangements (ESDI/ESDP), but until these arrangements solidify and until the EU accepts Turkey as a full partner in defense decisionmaking—a distant prospect at best—the defense link with the United States will remain paramount. A very troubling scenario, from the Turkish point of view, would be the rise of a more or less capable EU defense structure, outside of NATO, from which Turkey, as a non-EU member, is largely excluded. This would become a worst-case scenario for Turkey if this development were coupled with a waning of NATO's security role and a progressive disengagement of America from European defense. At base, Turkish views on ESDP are similar to those prevailing in both the Clinton and Bush administrations. Washington has been supportive of a relatively full role for Turkey in EU defense decisionmaking, consistent with the U.S. view that EU defense efforts should not discriminate against allies that are not members of the union. That said, the tough Turkish stance on ESDP matters has tested the limits of American support.

A second Turkish stake in the bilateral relationship turns on the pivotal role of the United States as a security arbiter in adjacent regions, that is, beyond the containment of Russian power. Turks often refer to their existence in a "dangerous neighborhood," with chronic instability on their borders. The containment of diverse security risks, from the Balkans to the Middle East, benefits considerably from cooperation with the United States, as in relations with Russia. Europe, even a Europe with growing ambitions in the foreign and security policy spheres, is unlikely to exert the same weight in regional affairs. To be sure, the American involvement in such areas as the Gulf can pose dilemmas for Ankara, and policy interests and approaches do not always coincide. On balance, however, Turkey benefits from the continued American military presence in adjacent regions.

In this context, Turkey's foreign and security policy establishment views the evolving American debate over overseas engagement with some anxiety. Ankara is used to measuring the health of the bilateral relationship in rather narrow terms, assessing Washington's interest in Turkey as a strategic partner and with an eye on questions of arms sales, textile quotas, and human rights policies. There is, however, a growing sense that substantial and continued American engagement in areas of critical interest to Turkey, whether in the Balkans or the Eastern Mediterranean, cannot be taken for granted. In this sense, Ankara shares the general European concern about preventing a decoupling of American and European security, and views the prospect of a reduced American role in peacekeeping in Bosnia or Kosovo with alarm. But Turkish concerns are more complex, because Turkey's direct security concerns go beyond the Balkans to Eurasia and the Middle East, and because Ankara sees the U.S. presence as an essential part of a credible Western security guarantee. This is particularly true of ballistic missile risks from Iran, Iraq, and Syria, where the United States (and Israel) are seen as the only security partners capable of providing Turkey with at least a minimally effective means of defense in the coming years.

Stability and reconstruction in the Balkans will, of course, be strongly affected by EU policies, but Washington still has enormous influence and has been a consistent advocate for a Turkish role in the region. In the Middle East, U.S. involvement provides a measure of reassurance against the worst outcomes in Turkey's relations with its neighbors, even if Ankara disapproves of the economic and political aspects of containment policies in the Gulf. Turkish views on the constructive engagement of Iran are far closer to those prevailing in Europe. In the event of a future Syrian-Israeli peace settlement, the United States would almost certainly play a key role in making sure Turkey's interests are protected, whether on water supply or in restrictions on Syrian military redeployments along the border with Turkey. Under conditions of crisis between Israel and the Palestinians, Ankara favors an active role for the United States in the Middle East peace process.

A third and long-standing Turkish interest concerns access to American military equipment, training, and defense-industrial cooperation. Turkey is in the midst of a major military modernization program—one that is likely to remain substantial even in the wake of economic difficulties. Important aspects of the modernization program anticipate the purchase of American equipment or U.S. source technology.[12] Throughout the Cold War and to the present day, the United States has been the leading supplier of defense goods and services to Turkey, a relationship that has persisted despite the end of formal security assistance and periodic crises over arms transfers.[13] Despite efforts at diversification, Turkey still conducts roughly 80 percent of its defense-industrial activity with the United States.[14] Large numbers of Turkish officers have trained in the United States and military-to-military habits of cooperation are strong, although Turkish military contacts with other NATO allies—and Israel—are becoming more frequent and may eventually dilute an outlook that has been heavily focused on the United States.

The Turkish military has a clear preference for American systems, but is troubled by the unpredictability of American, and especially congressional, attitudes toward sales to Turkey. The experience of the 1964 "Johnson letter" linking the American security guarantee vis-à-vis the Soviet Union to Turkey's policy on Cyprus, and the outright arms embargo following the 1974 conflict on the island, have had an enduring effect on Turkish perceptions. In recent years, disputes over the transfer of American frigates, attack helicopters, and other advanced weaponry have arisen against a background of congressional concern over Turkey's human rights situation and fear of fueling an arms race in the Aegean. At the height of bilateral tensions over these issues in the mid-1990s, many Turks came to believe that the country faced a *de facto* American arms embargo, and the coexistence of supportive and punitive policies emanating from Washington raised the question of whether Ankara was regarded as an ally or a rogue state.[15]

---

[12]Prominent bilateral defense-industrial projects in the 2000–2002 time frame include the planned procurement of attack helicopters and early warning aircraft, Turkish participation in the Joint Strike Fighter program, new naval helicopters, heavy tanks, and the pedestal-mounted Stinger SAM. Turkish Undersecretariat for Defense Industries, Briefing to American-Turkish Council, Washington, March 26, 2001.

[13]Defense sales and credits to Turkey are now arranged on a commercial basis.

[14]American-Turkish Council, unpublished paper and discussion, Washington, March 26, 2001.

[15]We are grateful to our former RAND colleague Zalmay Khalilzad for this formulation.

The waning of the battle against the PKK in Southeastern Turkey and the mood of rapprochement in Greek-Turkish relations have improved the arms transfer climate. But concerns persist on both sides. In response, Turkey has moved to diversify its sources of military goods and services. Israel has been the leading beneficiary, with important new purchases, upgrade contracts, and training arrangements. Turkey would like to purchase and participate in the production of Israel's Arrow missile defense system, although the existence of U.S.-source technology in this program makes Turkish access subject to American approval. Other defense contracts have gone to Russia and European vendors in recent years. Realistically, however, the bulk of Turkey's defense modernization over the next decades is likely to involve cooperation, and especially co-production arrangements, with the United States. Such arrangements, most notably the manufacture in Turkey of over 300 F-16 fighter aircraft, have contributed enormously to the country's military capability, technical capacity, and international prestige.

Fourth, Turkey continues to look to Washington for support on key Turkish objectives outside the defense realm. The United States has, for example, been a leading proponent of the Baku-Ceyhan pipeline to bring Caspian oil and gas to world markets via a terminal on Turkey's Mediterranean coast. The pipeline scheme remains a cherished objective for Ankara. It would bolster Turkey's regional influence and limit that of Russia and Iran, leading competitors for Caspian oil transport. It would encourage the economic and political independence of the Central Asian republics and bolster their links to Turkey. The environmental risks associated with vastly increased tanker traffic through the Bosporus would be reduced. And Turkey would stand to receive significant economic benefits from pipeline construction and transit fees.

Washington has been active at the diplomatic level in support of Baku-Ceyhan. But the official U.S. policy remains one of support for "multiple pipelines" (i.e., Turkish and Russian, but not Iranian), and financial backing and guarantees have not been forthcoming. Moreover, it is arguable that the Bush administration will be even less interested than its predecessor in providing subsidies to energy schemes that would normally rise or fall on the basis of commercial viability. The commercial viability of the Baku-Ceyhan pipeline has been an open question, with many economic and political variables, including the potential relaxation of sanctions on Iran and the opening of an Iranian route for Caspian oil. The discovery of large new reserves in the Caspian region, and the growing potential for gas exports alongside oil shipments, have improved the outlook for Baku-Ceyhan, and the scheme now stands a good chance of moving ahead to completion. Turkish officials have argued that the West as a whole—and the United States as an energy security guarantor in the Gulf—has an overriding geostrategic stake in diversifying energy routes and reducing global reliance on oil shipments through the Strait of Hormuz. From Ankara's perspective, more vigorous American support for the scheme—possibly including subsidies—will be essential. In the end, however, Turkey will probably need to make a substantial financial contribution to the project, and the prospects for this in the current economic climate are uncertain.

Another critical example of U.S. support for Turkish objectives concerns the EU and Turkey's role in European affairs. The United States has been a consistent advocate of Turkish integration in Europe, including membership in the EU, and Ankara has looked to Washington for support at critical junctures. Washington has made the argument for closer Turkish integration on strategic grounds, arguing that anchoring Turkey ever more fully in European institutions is necessary if Turkey's longer-term Western orientation is to be guaranteed. This view parallels that of Turkey's foreign policy establishment. Lobbying by senior American officials played a pivotal role in the European parliament's decision to approve the Turkish-EU Customs Union in 1996. It was almost certainly a factor in the EU's December 1999 decision to offer Turkey candidate status at the Helsinki summit.

The Helsinki outcome changes the context for American advocacy in important ways. Europe has never been comfortable with U.S. lobbying on behalf of Turkey, and some European allies have questioned the American standing in European decisionmaking on enlargement questions. With Turkey's candidacy, Turkish-EU relations have moved into a more highly structured and legalistic pattern, with fixed criteria and fewer opportunities for arguments on strategic grounds. Moreover, the prospects for Turkey in the EU will now depend far more heavily on reform decisions taken inside Turkey.

Effective American lobbying in this area is now more likely to take place in Ankara than in Brussels, and Turks may be less comfortable with this. If Turkey's candidacy fails to progress favorably—and there is every indication that it will be a slow and uncertain process at best—Ankara may look to Washington to help. There may be fewer opportunities to do so with effect.

On the specific issue of Turkey's place in EU defense arrangements, Washington has been supportive of the Turkish position—to a point. The two countries are similarly NATO-centric and share a strong interest in seeing European defense arrangements develop, to the extent possible, within a NATO framework. Turkey has looked to the United States for support in its argument that non-EU NATO members should be fully integrated in EU defense decisionmaking, especially where their interests are directly involved (and since many of the likely contingencies for a European rapid reaction force are in Turkey's neighborhood, Ankara's takes this question very seriously). American officials have been strong advocates for the Turkish view in NATO and other settings. At the same time, this issue has also been a point of friction in the bilateral relationship. Ankara's hard-line stance on ESDP, amid threats of a Turkish veto on NATO enlargement and other matters, has on several occasions threatened to drive the EU to develop its defense plans outside NATO—a development that complicates Washington's European and NATO policies.

The burgeoning Turkish-Israeli relationship has developed without active or direct U.S. support, but the American element is nonetheless present in the Turkish calculus. Ankara favors the emergence of more extensive trilateral cooperation, especially in the Eastern Mediterranean, and sees its already well-developed relationship with Israel as a basis for more ambitious political, defense, and investment initiatives. More important, many Turks have seen their relationship with Israel as a vehicle for improving Turkey's image and political position in Washington and, in particular, in Congress. To be sure, there has been some benefit along these lines, although in a less-direct fashion than many Turks might have wished. Supporters of Israel in Washington have been cautious in taking up the Turkish case, but cooperation among Turkish and Israeli advocacy groups in the United States has expanded.[16] The relationship with Israel has broadened interest in Turkey among those foreign policy experts whose principal frame of reference has been Middle Eastern affairs. It has also given government and nongovernment analysts a new frame for considering the Turkish role and Turkey's contribution to Western strategic objectives.

The Turkish private sector has been especially articulate in describing the opportunities for Turkish-American cooperation in support of Turkey's economic development and the country's role in regional commerce beyond pipeline issues. The opening of the Turkish economy under Özal, and the waning of the statist model that had shaped the Turkish economy for decades, created the basic conditions for increased bilateral trade and investment. Both expanded in the 1990s, although not at the pace seen in Turkish relations with other partners such as Russia, Israel, or even the EU.[17] Given Turkey's looming energy deficit, it is noteworthy that some of the most significant new American ventures in Turkey have been in the power generation field. Notwithstanding Turkey's designation as a "big emerging market" by the Department of Commerce, the economic dimension of the relationship has developed more slowly than many had anticipated. Weak intellectual property protection, the lack of acceptable arbitration procedures (now remedied), the slow pace of privatization, chronic high inflation and financial instability, and the perception of political risk have inhibited investment. Ankara, for its part, remains focused on raising American quotas for the import of Turkish textiles—a traditional mainstay of Turkish

---

[16]A good example is provided by the study visit to Israel and Turkey for American foreign policy experts organized jointly by the Assembly of Turkish-American Associations and the American Jewish Committee in January 2001.

[17]Between 1991 and 1999, for example, American exports to Turkey increased from roughly $2.5 billion to $3.2 billion per year (actually slightly higher in 1997). Turkish exports to the United States increased from roughly $1 billion to $2.6 billion in the same period. U.S. investments in Turkey have varied widely from year to year, a reflection of economic and political instability in Turkey, showing only modest increases over the decade. Figures cited in Abdullah Akyuz, "U.S.-Turkish Economic Relations at the Outset of the 21st Century," *Insight Turkey*, Vol. 2, No. 4, October–December 2000, p. 74.

trade with the United States, and reducing nontariff barriers to Turkish agricultural exports.[18]

The 2000–2002 financial crisis in Turkey is a reflection of fundamental economic and political problems evident for many years. But it has had a disastrous effect on foreign investment and will doubtless shape the perceptions of American investors and trading partners for some time to come. Such perceptions are not insurmountable given the appropriate remedies, as the Mexican case illustrates. Yet the current economic crisis calls into question the near-term prospects for bolstering economic ties between Turkey and the United States. It also makes the issue of American support for Turkey in international financial institutions even more critical and raises the question of the U.S. role in possible future bailouts of the Turkish economy.

After an initial period of hesitation in the early spring of 2001, the Bush administration threw its support behind a package of IMF and World Bank support for Turkey totaling some $17 billion. At the same time, as a condition for its support, Washington insisted on the implementation of a sweeping set of economic reforms and austerity measures. Having come to power with an avowed distaste for international economic "bailouts," and against the background of another financial crisis in Argentina, the administration's decision to support the financial rescue package was greeted with relief in Turkey. Nonetheless, many Turkish commentators have complained about the grudging nature of American support as well as accompanying pressure from Washington over issues as diverse as Cyprus, Iraq, the handling of Chechen sympathizers in Turkish custody, and constitutional reform. Critics in Turkey's nationalist circles and on the left have portrayed the financial package as a compromise of Turkish sovereignty engineered by the IMF under American direction, a suspicion encouraged by Kemal Dervish's long residence in Washington. These perceptions reflect the close linkage between Turkey's economic fate and U.S. policy preferences that exists in the eyes of many Turks.

---

[18]For a good discussion of these trade and investment concerns see Akyuz, "U.S.-Turkish Economic Relations at the Outset of the 21st Century," pp. 71–81.

CHANGING BILATERAL IMAGES

Beyond questions of national interest and strategic concerns, the bilateral relationship is also shaped by questions of affinity and familiarity.[19] In contrast to other transatlantic relationships, this dimension of Turkish-U.S. relations is not well developed. The Turkish-American community is relatively successful and affluent, and increasingly organized and active, but it is modest in size at perhaps 300,000–350,000.[20] The community lacks the weight of its counterparts from Greece and Armenia in the American foreign policy debate. This fact figures heavily in Turkish interpretations of American policy and goes some way to explain the Turkish focus on lobbies as a feature of American policymaking. Turks often assert that their country has no effective lobby in Washington. Others assert that Turkey has historically had a very effective lobby in the form of successive administrations with a strategic interest in Turkey. American specialists on Turkey, defense analysts, and defense industries have also been strong advocates for Turkish interests over the past decades.[21] It is perhaps more accurate to describe the Turkish position vis-à-vis its traditional lobbying opponents in Congress and elsewhere as asymmetrical rather than unfavorable *per se.*

Among those who follow Turkish affairs closely, the debate is often polarized between those focused on Turkey's geostrategic importance and those concerned with Turkey's problems of democratization and human rights. Given Turkey's position as a long-standing ally and member of NATO, it is remarkable that American specialists, and officials charged with the management of the bilateral relationship, often find Turkish society, and especially civil-military relations, perplexing. It is not unusual to hear senior officials describe their frustration in trying to understand how Turkey "really works." Others have described the process of making sense of Turkish policymaking as an exercise in "Kremlinology."

---

[19]The "familiarity" issue is discussed in Alan Makovsky, "Marching in Step, Mostly!" p. 37.

[20]Interview with Orhan Kaymakcalan, President of the Assembly of Turkish-American Associations, "Challenges to Turkish Identity in the U.S.: An Interview with the ATAA," *Insight Turkey*, Vol. 2, No. 4, October–December 2000, p. 81.

[21]See Abramowitz, "The Complexities of American Policymaking on Turkey," pp. 3–35.

Few Americans are familiar with Turkey, and even well-informed Americans tend to share a perception of Turkey as culturally and politically exotic. Outside foreign policy circles, where Turkey is often seen through a NATO lens, the image of the country is Middle Eastern rather than European. Ironically, Turkey's active role in the Gulf coalition reinforced the perception of Turkey as a valuable Middle Eastern ally. At the popular level, Turkey's image has never fully recovered from the popularity of *Midnight Express*, a film that many Turks believe set back the development of American tourism in Turkey for decades. The number of American visitors to Turkey has increased substantially in recent years, reaching some 500,000 in 2000, but it lags far behind tourism from Europe, which is a major revenue earner for the Turkish economy.[22] Tourism to Turkey is also highly dependent on a perception of regional stability, and crises in the Balkans or the Middle East, as well as international terrorist incidents, can have a highly damaging effect on American tourism, however unjustified. This effect was made clear by the drop in American tourism following the September 11, 2001, terrorist attacks and the subsequent intervention in Afghanistan.

Overall, the lack of accurate knowledge about Turkey and the relative paucity of "people-to-people" contacts mean that bilateral affinity is underdeveloped and is not commensurate with Turkey's importance in foreign policy terms. This reality also contrasts with the increasingly active specialist debate about Turkish affairs in the United States, and a marked increase in the coverage of Turkish topics in leading American newspapers in the second half of the 1990s.

Turkish images of the United States are also distinctive—and changing. Turkish analyses of this question normally point, quite correctly, to the anti-American instincts of both the nationalist right and the Turkish left. The former have tended to see ties to the United States as a threat to Turkish sovereignty. The latter have shared the Cold War tradition of concern about the American model of capitalism, American "imperialism," coupled with nationalist instincts of their own. These intellectual strains, common across Europe and the Middle East during the Cold War, still weigh heavily in the contem-

---

[22]In 1991, in a period of regional tension, only 79,000 American tourists visited Turkey. The figure was 182,000 in 1992, and reached 290,000 by 1995. Some 500,000 Americans visited Turkey in 2000.

porary Turkish debate. Even among elites, there is often a marked suspicion of American aims. Turkey's diplomats chafe at the need to adjust the country's foreign policy to meet American expectations. The military establishment values its strategic and technical collaboration with the United States, and has been a (perhaps *the*) leading interlocutor in bilateral relations, but it is especially sensitive to sovereignty questions, and these are often at the forefront of discussions with Washington on Incirlik and other matters.

Broadly speaking, the Turkish private sector—a relative newcomer to foreign policy debates in Turkey—is less suspicious and more positively inclined. Leading Turkish business groups have devoted considerable energy to promoting trade and investment ties to the United States, with the goal of augmenting and diversifying Turkey's Euro-centric economic relations. TUSIAD has opened a representative office in Washington with a public policy as well as a trade promotion mandate. The phenomenon of globalization—as fashionable a topic in Turkey as elsewhere—is often closely associated with the United States in Turkish perceptions. Here, too, the business community tends to be more comfortable with the notion of globalization than those in political circles, on the right and the left, and more favorably disposed toward the American role.[23] It is notable that in Turkey's economic crisis of 2001, lobbying for support in Washington has been a priority for Turkish business groups. Overall, the private sector has emerged as a far more prominent interlocutor in bilateral relations and now plays a large role in shaping American attitudes toward the country. Turkey's economic troubles, and the focus on corruption and mismanagement in public-private sector relations in Turkey, have caused special dismay in Washington, where the "dynamism" of Turkey's private sector has been a feature of most discussions about the Turkish scene in recent years.

The emergence of a more diverse foreign policy debate in Turkey, with new elites participating, has contributed to a more positive perception of the United States. Despite some important policy differences, the general tone of the relationship has arguably never been better. In the view of many Turkish observers, President Clinton's

---

[23]See Lesser, *Strong States, Difficult Choices: Mediterranean Perspectives on Integration and Globalization.*

November 1999 visit to Turkey in the context of the OSCE summit marked a high point in the relationship. Turks were impressed by the substance and tone of key speeches, including a speech to the "not-so-pro-American" Turkish parliament that was greeted with great enthusiasm by Islamist deputies, among others.[24] Turkey's Islamists have, as a rule, been more concerned with domestic than foreign policy, and their perspectives on the United States and the bilateral relationship are far from uniform. Although sharing some of the nationalist concerns of the secular right (e.g., about Cyprus) that encourage a wary view of the United States, many Islamists also see themselves as potential beneficiaries of American pressure over human rights and democratization. Turkey's mainstream Islamists, including leading members of the now banned Virtue Party, have generally been eager to engage American policymakers and observers. Turkey's Islamist parties have generally expressed shock over the September 2001 terrorist attacks on the United States, although they have been less supportive of the idea of Turkish military contributions to operations in Afghanistan or Iraq.

There can be little doubt that most Turks know more about the United States than their American counterparts know about Turkey. The pervasiveness of American culture, business, and media and the prominence of American actions worldwide assure that this is so. Nonetheless, even well-informed Turks are often puzzled by the workings of the policy process in Washington. Decades of battles with Greek, Armenian, human rights and other groups in Washington have encouraged the view that Turkish-American relations are influenced, above all, by the clash of lobbies. The engagement of American policymakers on substantive issues and the enhancement of programs to address the "affinity deficit" at the public level appear to receive less attention. This reality is particularly meaningful at a time of strategic flux in which key international relationships are being redefined and Turkey, even as an important regional ally, must compete for the attention of American policymakers.

---

[24]Candar, "Some Turkish Perspectives on the United States and American Policy Toward Turkey," in Abramowitz, *Turkey's Transformation and American Policy*, p. 147.

LOOKING AHEAD

After decades in which the contours of Turkey's relations with the United States were well established, the bilateral relationship at the start of the 21st century faces many sources of change. The progressive transformation and modernization of Turkish society, with all the associated stresses and strains, may be the most significant element in this equation. Changes on the domestic scene have brought new issues to the fore and could ultimately reduce or eliminate many long-standing obstacles to an even closer relationship, including shortcomings on human rights and political and economic reform.

The political and social dimensions of Turkey's economic travails will also be meaningful to Washington. A chaotic and less-prosperous Turkey will have little energy to play a positive regional role and little ability to fulfill its promise as a "big emerging market"—themes that have been central to official visions for the bilateral relationship. The pace and character of change on the domestic scene will be a determinant of Turkey's progress in relations with the EU—an essential backdrop for the future of Turkey's relations with the United States. Domestic developments will also shape the way Turks view broader questions of globalization in which the United States looms large. Turkey's internal transformation is likely to be a key, enabling factor setting the tone and limits of the bilateral relationship as seen from Washington. Similarly, for Turks, the American response to Turkey's internal challenges, including the country's economic crisis, will be a key test of the health of the relationship and the atmosphere for cooperation on other issues.

At a fundamental level, Turkish and American interests are broadly convergent. Both states are inherently status quo powers with respect to the regional and international environment. Despite increasing activism in key areas such as the Middle East, Turkish foreign policy can still be characterized as cautious and conservative in overall terms. Both countries are, for different reasons, relatively security conscious, and the bilateral relationship retains immense value as seen from Ankara. This shared security consciousness is likely to be strongly reinforced by the new primacy of counterterrorism in U.S. strategy. Nonetheless, at the level of policy approaches, there are some important areas of ongoing and potential divergence.

Looking ahead, several issues stand out as sources of change—both challenges and opportunities—for the bilateral relationship.[25]

First, regardless of the outcome of Turkey's EU candidacy, the European factor is likely to be a leading influence on the bilateral relationship in the 21st century. The longer-term implications of a more European Turkey in policy terms may be significant. The net result is likely to be greater normalization and maturity in relations between Washington and Ankara, as has been the pattern elsewhere across Southern Europe. If, by contrast, the Turkish-EU relationship stalls or deteriorates (e.g., over the question of Cypriot accession, lack of reform in Turkey, or ESDP), there will be greater reliance—and pressure—on Turkey's relationship with Washington. This could prove an uncomfortable reality for both sides, particularly against a backdrop of tension in transatlantic relations.

Second, the advent of the Bush administration in Washington places the question of regional policies in sharper relief. Turkey and the United States may seek peace and stability in areas of shared concern, but policies differ. Iran, and above all Iraq, will be key questions in this regard. A tougher American stance in the Gulf, and especially a renewed military confrontation with Iraq or an effort to tighten sanctions, would be met with dismay in Ankara. It could also prove a test of Turkish solidarity in Washington. Against this background, both countries face decisions about tangible matters such as the use of Incirlik air base after the eventual end of Operation Northern Watch and the conduct of Turkish-Iraqi trade within and outside the UN sanctions regime. From a Turkish perspective, the best outcome might well be continued military containment of Iraq, accompanied by a loosening of the sanctions regime—and no support for opposition movements in Iraq that might spur chaos and Kurdish separatism in the region. American policy preferences make this unlikely. In other areas such as Central Asia, the Balkans, and the Aegean, bilateral perspective are more congruent. Washington has been a strong supporter of Turkey's prospective leadership of peacekeeping operations in Afghanistan (ISAF) and has agreed to help defray the costs of Turkish participation. In the Middle East

---

[25]For a perspective on some of these issues, see Alan Makovsky, "Turkey and the Bush Administration: The Question Marks," *Policywatch* No. 527, Washington, D.C.: The Washington Institute for Near East Policy, March 30, 2001.

peace process, Ankara will likely support an engaged rather than an arm's length approach from Washington. The strong Turkish-Israeli relationship only increases Ankara's stake in this area.

Third, a decade after the end of the Cold War, policy toward Russia is again at the forefront. A deterioration of U.S.-Russian relations will increase risks for Turkey in the Caucasus and Black Sea, as well as in the Middle East and the Balkans where Russian policies are meaningful. At the same time, Turkish concerns about Russia mirror those in the United States, and the American connection will remain the cornerstone of Turkey's deterrent posture vis-à-vis Russia. For these reasons, dialogue on the management of relations with Moscow should be a prominent item on the bilateral agenda for the future.

Fourth, energy policy is likely to become an even more important part of the relationship. The elements here include Turkey's own energy needs, America's growing interest in energy security issues, and Turkey's role as a conduit for Middle Eastern and Caspian oil and gas. Much attention has been focused on the prospects for the Baku-Ceyhan pipeline. Many Turks see continued, even enhanced, American support as critical to the outcome. But the American enthusiasm for subsidizing the project is very limited. Even apart from Baku-Ceyhan, the restoration of full Iraqi oil exports via Turkish pipelines (the capacity of these existing lines is roughly twice that of the proposed Baku-Ceyhan route) would strongly reinforce Turkey's role in the world energy picture.

Fifth, the rationale for a "strategic" relationship will go beyond Turkey's geographic position in relation to regions of shared concern. Many of the most prominent foreign and security policy problems in the new environment are transregional. A prominent example is the challenge of missile proliferation. Ankara's perspective on this issue is perhaps closest to that of the United States, and Turkey's interest in missile defenses is correspondingly strong. Turkish-U.S. dialogue and cooperation on missile defense could emerge as an important subset of the increasingly energetic transatlantic debate on this topic. If a regionally based or "boost-phase" missile defense architecture is pursued, parts of this system might well be based in Turkey, at which point this will move from a conceptual to a practical issue in the bilateral relationship.

Finally, policymakers in Ankara and Washington will continue to seek, with some success, a more diverse relationship featuring increased economic and other forms of cooperation outside the security realm. But the primacy of security issues in Turkish-U.S. relations is likely to endure for structural reasons. These reasons include the flavor of Turkish and American policy concerns, persistent instability in adjacent regions, the impetus of decades of security cooperation, uncertainties regarding Russia, and, not least, the existence of other more natural economic partners for Turkish business. It is notable that in the midst of Turkey's economic crisis, Turkey's advocates—including the Turkish private sector—have made the case for support in strategic rather than economic terms.

The future bilateral relationship will need to reflect a changing Turkey, a changing strategic environment, and an evolving foreign policy debate in the United States. It must also accommodate the more rigorous measurement of Turkish national interests that has accompanied the country's more active external policy and growing regional power. In all likelihood, the relationship will be more diverse, within limits, and involve a wider range of interlocutors. More than ever before, the character of Turkish-U.S. relations will depend on external variables, such as Turkey's relationship with Europe and the evolution of Russia, outside the bilateral agenda narrowly defined.

Chapter Eight

CONCLUSION

In the last decade Turkey has emerged as a more active and important actor on the international stage. After the collapse of the Soviet Union, Turkey rediscovered a world of interests and affinities stretching "from the Balkans to Western China"—areas that had been largely absent from the mainstream Turkish foreign policy debate, not just since the start of the Cold War but since the foundation of the Republic. More recently, analysts have focused on the increasing activism in Turkish external policy. With few exceptions, this activism has been evident largely in traditional areas of interest such as Europe, as well as areas of perceived risk, above all the Middle East.

Turkey is, at base, a conservative society with a conservative approach to public policy in most spheres. Almost 80 years after the founding of the Turkish Republic this remains true. However, Turkey today is in the midst of a period of important political change that could have a profound effect on its foreign policy evolution. Looking ahead, several significant, open questions will shape the Turkish foreign and security policy debate. They will also shape to a considerable extent the character of Turkish relations with the West. These key questions concern the future shape of Turkey as a society, Turkey's international identity, its regional behavior, and its place in a globalized world.

WHITHER TURKEY?

Turkey's current economic travails only serve to underscore the importance of the country's internal evolution in determining what is possible and what is likely in Turkish external policy. Turkey is truly

at a crossroads. After more than a decade of substantial modernization, Turkey faces a crisis of leadership and reform. Few Turks, outside of the most traditional government circles in Ankara, would dis-
agree with the notion that the process of change in Turkey has
reached an impasse that requires some dramatic changes in the way
the country is governed, and by whom. Visions of what should come
next differ at the level of orientation but not scale. Incrementalism
and "muddling through"—approaches that have characterized
Turkish policy in the past—are unlikely to be sufficient in the future.
The threshold for social unrest in Turkey remains high, but a continuing economic crisis, with social and political cleavages left unresolved, could push Turkey toward instability, making more extreme
or chaotic outcomes a possibility. The foreign policy consequences
would be substantial. In particular, a Turkey in turmoil would likely
find Europe even more resistant to the idea of Turkish membership
in the EU.

The argument about Turkey as a pivotal state turns on the potential
for the country to affect a wider area through its foreign policy
behavior but also through internal developments. The notion of
Turkey as a regional model reflects the positive side of this potential.
But an impoverished and unstable Turkey would have very different
and very negative consequences for Europe, Eurasia, and the Middle
East. An unstable hinterland might impede reconstruction and integration in Southeastern Europe. It might reinforce an existing tendency toward instability in the Caucasus and Central Asia. Security
perceptions in Athens and Moscow would be affected. An unstable
and unpredictable Turkey would contribute to a deteriorating security environment in the Middle East and would limit Western options
in the Gulf and elsewhere. From a NATO and an EU perspective, a
troubled Turkey makes it more likely that Ankara will be a "con-
sumer" rather than a "producer" of security on the European
periphery.

The scenario of a troubled, inward-looking, and more nationalistic
Turkey, with a more limited but also less predictable foreign policy, is
not the most likely case. But it is a possibility. More plausibly,
Turkey's resources and resilience—and the climate of incipient
change—will eventually lead to significant reforms on the political as
well as the economic scenes. The result in this case is likely to be a
more modern and stable Turkey, better integrated in European insti-

tutions, more comfortable with the challenges of globalization, and more moderate and multilateral in its foreign policy outlook. This is a scenario that would benefit Turks, Turkey's neighbors, and Turkey's allies.

As a former imperial power, Turks are used to thinking strategically, and the modern strategist finds no shortage of plausible and implausible theories in the Turkish policy debate. The economic crisis and the natural focus on domestic politics have, however, reduced strategic thinking to a minimum. For the moment, the country's intellectual and political energies are focused elsewhere, with an emphasis on immediate domestic challenges. But the resolution of these internal challenges will have a critical effect on Turkey's future geopolitical role. If Turkey proves unable to overcome its domestic difficulties, its ability to play an active and constructive role in international affairs will be sharply reduced.

WHERE DOES TURKEY FIT?

It is arguable that modern Turkey has functioned as part of several systems—European, Middle Eastern, and Eurasian—while remaining on the cultural and political periphery of each. At an important level, the Turkish foreign policy debate is a constantly renewed argument about identity. There is a tendency among many highly Westernized Turks to regard discussions about Turkish identity with distaste. Surely Turkey's EU candidacy means that the question of identity has been resolved? But this question of identity has not been resolved and will in all likelihood remain open for the foreseeable future. There is nothing pejorative or unnatural about this. Europe itself is witnessing an extended debate about its own identity as it contemplates enlargement, monetary union, and a more common approach to foreign policy and defense. Turks rightly sense that the EU's approach to the question of Turkey's (and other nonmembers') role in ESDP is not simply institutional in nature. It includes an identity dimension: Who is "European" and who is not? Where do Europe's frontiers end? Who and what are we defending?

Helsinki summit decisions notwithstanding, the issue of where Turkey fits in a changing Europe remains unresolved on both sides. Turkey's own internal evolution and convergence with European norms is a fundamental, enabling condition for the promise of European membership to be fulfilled. An evolving Europe, possibly with multiple circles and "variable geometry," could become more comfortable with the challenges of scale and identity associated with Turkish membership. On the other hand, the elements contributing to mutual ambivalence about membership are unlikely to be resolved easily. Europe may continue to hold Turkey at arm's length even against a background of growing Turkish integration and convergence with Europe in a political, economic, and social context. Or, the failure to meet key aspects of the EU's Copenhagen criteria, or lack of progress on Cyprus and the Aegean, could relegate Turkey to a hollow candidacy. In foreign policy terms, the question of Turkey's European integration is full of gray areas. It is perfectly plausible that Turkey can remain, as it always has, a functioning part of the European system short of full EU membership. Absent Cold War conditions, however, it may be increasingly difficult for Turkey to play an effective role in the Euro-Atlantic system without progress on the domestic reform agenda.

To what extent might alternative Middle Eastern, Muslim, or Turkic identities augment or replace a European identity that remains unconsolidated? The short answer is that this is most unlikely. Over the next decade, Turkey might well find itself with more active economic and political ties to Eurasia and the Middle East. This could be the product of a more dynamic Russia or the full reintegration of Iraq and Iran in the international system. Turkey is a potential beneficiary of both possibilities. Or it could be the product of political developments inside Turkey, including the emergence of a reformist-religious or a religious-nationalist synthesis, with fewer reservations about ties to the Muslim and Turkic worlds. It is nonetheless difficult to imagine the practical basis for such reinforced ties replacing the strategic (in the sense of comprehensive political, economic, and defense interests) relationship with the West. Europe is likely to remain the overwhelmingly important economic partner for Turkey, and short of a U.S. retreat from engagement in Europe and the Middle East, Washington will remain Ankara's key security partner. As a matter of identity, one possibility for the future is greater acceptance of a more balanced orientation between East and West, as the Kemalist tradition becomes more diffuse.

HOW WILL TURKEY ACT?

The heightened activism and assertiveness in Turkish external policy in recent years should prove durable. Immediate economic challenges may well leave Turkey with less energy and fewer resources to devote to foreign policy in the short term. But the bases of a more active approach, from the rediscovery of regional interests to the expansion of the public debate on foreign policy questions, are likely to endure and make themselves felt in the future. Indeed, one of the consequences of Turkey's recent travails may be a heightened sense of national interest, especially as it relates to trade and economic development. And ultimately, a more prosperous and integrated Turkey may see new reasons for active engagement in adjacent regions, alongside the country's international partners.

Turks continue to stress the insecurity of their neighborhood as a rationale for cautious engagement, and sometimes intervention. At the opening of the 21st century, Turkey's neighborhood remains extraordinarily troubled. The resulting regional challenges will be difficult, perhaps impossible, for Ankara to ignore. In Southeastern Europe, the process of political and economic reconstruction is likely to be prolonged, with the potential for further destructive conflict, especially in the Southern Balkans. Short of a serious break in Ankara's relations with the West, Turkey will remain a conservative, multilateral, and significant actor in the Balkans. But Ankara's policy is likely to remain in step with Western policy, not run counter to it.

With Athens, Ankara shares a strategic stake in the resolution of Aegean disputes. Greece and Turkey have shown an ability to act in concert in the Balkans and have recently moved toward a more flexible and less risk-prone stance in their bilateral relations. However, Ankara's future willingness to consolidate Turkish-Greek détente and, in particular, to help resolve the Cyprus problem will turn critically on the overall character of relations between Turkey and Europe. A Turkey disillusioned or bitter in its relations with the EU will have fewer incentives to compromise on the Aegean or Cyprus. The result could be a return to brinkmanship in Greek-Turkish relations and an additional burden on European and American diplomacy in the Eastern Mediterranean.

Turkey's rediscovery of a larger Turkic world in the Caucasus and Central Asia has had a significant effect on Turkey's perception of its national interests. It has also been a vehicle for more active involvement, officially and by the Turkish private sector. Initial expectations about the scope of Turkey's role have proven to be inflated. But the opening of former Soviet areas that had been off-limits, intellectually, politically, and in practical terms, has had enduring consequences for Turkish foreign policy. Opportunities in Eurasia have not replaced more pressing interests in the West, but they have placed these and other interests in a different perspective.

If Turkey is more central in European, American, and Russian calculations today, this is so in large part because the field for Turkish external policy is now much broader—and many of the new possibilities lie in Eurasia. Continued economic growth in Turkey will require access to new sources of oil and gas, from Russia and the Caspian. A more complex web of oil and gas pipelines will make these energy links a "permanently operating factor" in Turkey's foreign policy. With the Baku-Ceyhan pipeline plan, Turkey is a leading competitor in Caspian geopolitics. But perhaps more significantly, Turkey is now part of a complex and highly interdependent system of energy supply, and shared economic interests, spanning Eurasia. Even without Baku-Ceyhan, Turkey is emerging as a key entrepôt and transit state for energy supplies headed to Europe from Eurasia and the Southeastern Mediterranean.

Turkey's ability to play a more active role in the energy field, however, will be significantly affected by political developments in the Caucasus. Since the mid-1990s, Turkey has succeeded in enhancing its role as an important regional actor in the Caucasus, strengthening ties to both Azerbaijan and Georgia. But the political situation in the Caucasus is extremely fluid. Both Azerbaijan and Georgia will face succession issues in the near future. How these are resolved could have a significant effect on Turkey's interests in the Caucasus, especially the prospects for completion of the Baku-Ceyhan pipeline.

These new factors in the Turkish foreign policy calculus will have the effect of reinforcing a very traditional Turkish foreign and security policy concern about Russia. For centuries, Turkey was at the center of Russian-Western interaction in security terms. In the post–Cold War climate, the consequences of alternative paths in Russian-

Western relations are once again likely to be felt most directly on the European periphery, with direct implications for Turkey. Ankara will continue to have a strong stake in cooperative relations with a stable and satisfied Russia. A more chaotic Russian state, with turmoil on its borders, would create political and security vacuums around the Black Sea and the Caucasus, increasing the likelihood of refugee movements and violent spillovers affecting Turkey. Moreover, turmoil along Russia's south could stimulate nationalist sentiment inside Turkey and might lead to pressure for a more interventionist policy in Ankara. A resurgent and assertive Russia would similarly find more room for maneuver in peripheral areas, adjacent to Turkey, from the Balkans to the Caucasus. In short, the risks in Turkish-Russian relations are high over the longer term, and the need for deterrence and reassurance vis-à-vis Moscow will continue to drive a cautious and Western-oriented approach in Ankara.

Turks will continue to see the Middle East as an area of risk requiring an active, security-driven set of policies. The core of this activism is likely to remain the close connection between internal security issues—Kurdish separatism and Islamism—and the situation to Turkey's south and east. Developments inside Turkey, such as a gradual resolution of the Kurdish problem or the emergence of a less provocative relationship between religion and secularism in Turkish politics, could reduce the prominence of this connection, but it is likely to remain a factor of some weight in the Turkish foreign policy calculus. Thus, Ankara will prefer a cohesive and reintegrated Iraq that can be held at arm's length in security terms. Roughly the same approach will apply in relations with Iran—a policy orientation closer to that of Europe than that of the United States. If Turkey has a containment strategy in the Middle East, it will continue to apply, above all, to Syria, where the sources of bilateral friction are multiple and pronounced.

The Middle East has been the principal theater of Turkish regional activism in recent years. It is also the area where Ankara is willing and able to pursue a more assertive and unilateral set of policies. The task of balancing the defense of Turkish interests in the Middle East, sometimes with the use of force, without becoming embroiled in costly, strategic confrontations may be more difficult in the future for several reasons. Economic stringency is unlikely to derail the longer-term evolution of Turkey as a modern and highly capable power, capable of the projection of military force at some distance from its own Middle Eastern borders. In future Middle Eastern crises, Turkey will have the potential for significant intervention in its own right, in addition to facilitating Western power projection. At the same time, the spread of weapons of mass destruction and ballistic missiles across the Middle East will lead to far more credible threats of retaliation against Turkish territory. With the deterioration in Israeli-Palestinian relations again posing the possibility of a regional conflict, and with the question of Iraqi and Iranian futures unresolved, the ingredients for a much more challenging environment are in place as seen from Ankara. Here, as in other regional settings, the risks to Turkey of an activist stance would probably be far greater in the absence of a predictable Turkish security relationship with the West.

The new emphasis on combating terrorism in the wake of the September 11, 2001, attack on the United States is likely to complicate Turkish foreign policy, especially in the Middle East and Central Asia. As a Muslim country, Turkey will want to ensure that the war on terrorism does not become a "civilizational" struggle between Islam and the West. Ankara will also have to weigh its interest in supporting U.S.-led actions against terrorism with its own national interests in the Middle East and the Persian Gulf.

There is also some uncertainty regarding the effect of these developments on Europe. European attitudes toward Turkish membership in the European Union remain ambivalent despite the Helsinki decision. The war on terrorism may reinforce this ambivalence as Europeans become even more reluctant to see a Muslim country enter the EU, fearing the import of "Middle Eastern" conflicts. This could further complicate Turkey's prospects for EU membership, leaving Turkey increasingly frustrated and disappointed.

In short, Turkey's ability to play a strong regional role is likely to face increasing challenges in the coming decade. Turkey may well be a more confident and capable foreign and security policy actor in the 21st century, but it will face many new risks if this activism is played out in a unilateral context. The need to avoid acting as a "lone wolf" gives Turkey an interest in maintaining a diverse set of security relationships—transatlantic, European, Israeli, possibly even Russian under favorable conditions. The effectiveness of these ties will be a key variable in Turkey's regional influence.

WHAT PLACE IN A GLOBALIZED WORLD?

The diversity of challenges and opportunities on Turkey's borders encourages Turks and others to see Turkish policy through a regional lens or, more accurately, a set of regional lenses. But as the unresolved question of Turkish identity suggests, Turkish policy has always had a wider systemic context. As our discussion of domestic developments suggests, a more liberal, outward-looking Turkey is one quite likely path, but it is not the only one. Turkey could well experience a period of retrenchment in which the country turns inward and acts with a more nationalistic flavor internationally.

The costs and consequences of a retreat from integration and globalization are probably increasing. By many measures, from tourism and foreign remittances to telecommunications and Internet usage, Turkey is already a highly globalized country. Turkey's urbanized elites are relatively comfortable with the phenomenon of globalization, and many of Turkey's most successful businesses have far-flung international interests. The prosperity of the last 20 years has been closely tied to the opening of the economy and the progressive globalization of Turkish society at many levels.

Nonetheless, Turks are keenly aware of the competitive pressures and vulnerabilities that integration and globalization imply. The current economic crisis has only served to reinforce a long-standing sensitivity about globalization in light of Turkey's long tradition of state involvement in virtually all aspects of life and a very highly developed notion of national sovereignty. It is increasingly clear that Turkey cannot preserve the traditional prerogatives of the Kemalist state if it wishes to integrate more closely with Europe and participate more effectively in a globalized system. More precisely, Turkey *could* hold to traditional ideas of state sovereignty—and many Turks may favor this—but it will pay a high price to do so. Moreover, with political and economic questions becoming more central to Euro-Atlantic relations, it is likely that a more sovereignty-conscious and inward-looking Turkey would find its security relationships troubled as well. If economic and political reform fails in Turkey, many Turks may find it convenient to blame international financial institutions and the phenomenon of globalization more generally. A climate of resentment would inevitably affect the quality of Turkey's foreign and defense policy cooperation with the West.

As Turkish leaders have discovered in previous periods of political and economic turmoil, internal problems can have an isolating effect, chilling the climate for foreign investment and diplomatic cooperation with allies. In the absence of Cold War strategic imperatives, the link between internal stability and international engagement in and with Turkey may be even closer. Turkish default on its international debt, unlikely but not impossible, would have a devastating effect on the country's international standing and would reinforce an existing tendency in some quarters, especially in Europe, to see Turkey as "part of the problem." Further large-scale financial assistance to forestall a Turkish default would probably imply further draconian conditions and an insistence on fundamental political as well as economic change. Ultimately, Turkish policymakers may count on the idea that their country is simply too important in geopolitical terms for Europe and the United States to ignore—that the systemic implications of a Turkish collapse will always compel international intervention. They may be right. But Turkish-Western relations will not benefit from too many such test cases of the country's pivotal status.

Some aspects of globalization may imply an end of geography. For Turkey, the rise of political, economic, and military issues that cut across traditional regional lines and span Europe, Eurasia, and the Middle East will be central to perceptions of the country's geopolitical importance. Geography makes Turkey a key partner in addressing transregional risks, from drug smuggling and refugee flows to terrorism and the proliferation of destructive, longer-range weaponry. It will also make Turkey an essential partner in capturing new diplomatic and commercial opportunities, whether through new lines of communication for energy or in new approaches to the Middle East peace process.

IMPLICATIONS FOR TURKEY'S WESTERN PARTNERS

These conclusions suggest a number of implications for the United States and Europe in their relations with Ankara over the next decade. First, Turkey's internal evolution is likely to be the leading determinant of the country's foreign policy potential and direction. The key choices in this regard will be made by Turkey, and the conditions for successful change will emanate largely from Ankara. Turkey must choose the pace and extent of reform in key areas, including the economy, the rules governing political parties, and human rights. The outcome will be critical to Turkey's relations with Europe and the United States. However, although the longer-term direction for Turkish society will be determined largely by domestic realities, Western assistance, both political and economic, can help to reduce near-term risks and enable Turkey to manage its internal problems more successfully. However, this support should be conditioned on Turkey's willingness to proceed with a coherent and sustainable reform program.

Economic and political reforms could well bring previously marginal political forces into policymaking positions. As a result, Turkey's partners will face the challenge of developing an effective dialogue with a wider range of forces, including the new breed of Islamists and Turkish nationalists, in addition to the traditional centrist and secular elites. The foreign policy inclinations of these elements are in flux, and early dialogue could help to ensure more moderate policies in future years.

Moreover, the Turkish political establishment faces a major generational turnover. Many of the dominant political figures of the last several decades—Ecevit, Demirel, Erbakan—are in their 70s and will soon depart from the political scene. They will be replaced by a new generation of Turkish leaders who may adopt quite different approaches to many problems than the older generation of leaders. As this transition unfolds—and this is likely to happen rather rapidly— Turkey may go through a difficult period of change. Thus, the West should begin now to establish close contacts with the new generation of leaders who will shape Turkey's future.

Second, Turkey's continued integration and convergence with Europe will be the leading external determinant of Turkish foreign policy behavior in the coming years. Helsinki established a path toward integration, but the outlook for Turkish membership remains highly uncertain. Turkish estrangement from Europe—a real risk—would have very negative consequences for Turkish regional policy in the Eastern Mediterranean and the Balkans and would render Turkey a far less-predictable security partner in Eurasia and the Middle East. A difficult relationship with Europe could open the way for a more nationalistic and unilateral Turkish foreign policy. By contrast, a stable, positive evolution of Turkish-EU relations would encourage a more predictable and multilateral approach across the board. It would also simplify and strengthen U.S.-Turkish relations, provided transatlantic relations as a whole remain cooperative and Washington remains engaged in European and Middle Eastern affairs. A situation in which Turkey is forced to choose between Europe and the United States, in an unstable strategic environment, would pose nightmarish dilemmas for Turkish policymakers and could create serious tensions in U.S.-European relations.

It is important, therefore, that U.S. and European approaches toward Turkey be in harmony. EU policy will have a significant effect on Turkey's future evolution. But given Ankara's strong security ties to the United States and Washington's security interests in the Eastern Mediterranean and Middle East, the United States will continue to be an important influence on Turkish policy. Thus, to avoid the emergence of new transatlantic differences over Turkey, U.S. and European policy needs to be closely coordinated.

This is particularly true regarding Cyprus. If mishandled, the accession of Cyprus into the EU could lead to a serious deterioration of Turkey's relations with the EU—and even stimulate a nationalist backlash toward the West more broadly. The current détente between Greece and Turkey could also be jeopardized, possibly leading to a new period of confrontation between the two countries. Both the EU and the United States have a strong stake in preventing such a development. Thus, U.S. and European policymakers need to give higher priority to achieving a Cyprus settlement.

Third, even under the most favorable conditions and with a multilateral orientation in Ankara, Turkish cooperation in regional affairs, including Gulf security, cannot be taken for granted. The trend toward more careful measurement of Turkish interests, and the willingness to act forcefully in defense of security objectives, is unlikely to weaken. To the extent that Turkey recovers from its economic and political difficulties, the coming years are likely to see a new Turkish debate about foreign and security policy in which traditional assumptions about the rationale for cooperation will be reassessed.

Thus, the use of Turkish assets by the United States and NATO in the future cannot be automatically taken for granted.

Finally, expectations regarding Turkey's international role should be tempered with a degree of realism. Turkey may well emerge as a more potent regional power in political, military, and commercial terms. But it faces strategic challenges in Eurasia and the Middle East that cannot be addressed with reference to national means alone. Effective Alliance relationships will be essential to an effective Turkish foreign policy, and these will impose their own constraints, particularly in an era of more dilute and conditional security ties. More broadly, Turkey's long-standing dilemmas regarding identity and the country's role in various international "systems" are unlikely ever to be fully resolved—nor is such resolution necessary. Key elements of the Kemalist tradition may fade—or be modified—but Atatürk's legacy is likely to continue to exert an important influence on Turkey's political evolution and differentiate it in important ways from that of other European states.

BIBLIOGRAPHY

Abramowitz, Morton, ed., *Turkey's Transformation and American Policy*, New York: The Century Foundation Press, 2000.

_____, "The Complexities of American Policymaking on Turkey," *Insight Turkey*, Vol. 2, No. 4, October–December 2000, pp. 3–35.

Ahrari, M. E., *The New Great Game in Muslim Central Asia*, McNair Paper 47, Washington, D.C.: Institute for National Strategic Studies, January 1996.

Akinci, Ugur, "The Welfare Party's Municipal Track Record: Evaluating Islamist Activism in Turkey," *The Middle East Journal*, Vol. 53, No. 1, Winter 1999, pp. 75–94.

Akyuz, Abdullah, "U.S.-Turkish Economic Relations at the Outset of the 21st Century," *Insight Turkey*, Vol. 2, No. 4, October–December 2000, pp. 71–81.

Alstadt, Audrey, *The Azerbaijani Turks: Power and Identity Under Russian Rule*, Stanford, CA: Hoover Institution Press, 1992.

Altunisik, Meliha Benli, "Turkish Policy Toward Israel," in Alan Makovsky and Sabri Sayari, eds., *Turkey's New World: Changing Dynamics in Turkish Foreign Policy*, Washington, D.C.: The Washington Institute for Near East Policy, 2000, pp. 59–73.

Aras, Bülent, "The Impact of the Palestinian-Israeli Peace Process in Turkish Foreign Policy," *Journal of South Asian and Middle Eastern Studies*, Vol. 20, No. 2, 1997, pp. 49–72.

_____, "Post Cold War Realities: Israel's Strategy in Azerbaijan and Central Asia," *Middle East Policy*, Vol. 5, No. 4, January 1998, pp. 69–70.

_____, "The Caspian Region and Middle East Security," *The Mediterranean Quarterly*, Vol. 13, No. 1, Winter 2002, pp. 86–108.

Athanassopoulou, Ekavi, *Turkey-Anglo-American Security Interests 1945–1952: The First Enlargement of NATO*, London: Frank Cass, 1999.

_____, *Israeli-Turkish Security Ties: Regional Reactions*, Jerusalem: Harry S. Truman Research Institute for the Advancement of Peace, Hebrew University of Jerusalem, March 2001.

Axt, Heinz-Jürgen, "Der Ägäis Streit—ein unlösbarer griechisch türkischer Konflikt?" *Südosteuropa Mitteilungen*, Nr. 2, 1999, pp. 137–151.

Ayata, Sencer, "Patronage Party and the State: The Politicization of Islam in Turkey," *The Middle East Journal*, Vol. 50, No. 1, Winter 1996, pp. 40–56.

Aydin, Mustafa, "Turkish Foreign Policy Towards Central Asia and the Caucasus: Continuity and Change," *Private View*, No. 9, Autumn 2000, pp. 36–44.

Aykan, Mahmut Bali, "The Palestinian Question in Turkish Foreign Policy from the 1950s to the 1990s," *International Journal of Middle East Studies*, Vol. 25, No. 1, 1993, pp. 91–110.

_____, "Turkey's Policy in Northern Iraq 1991–1995," *Middle Eastern Studies*, Vol. 32, No. 4, 1996, pp. 343–366.

Baev, Pavel, *Russia Refocuses its Policies in the Southern Caucasus*, Working Paper Series No. 1, Caspian Studies Program, Harvard University, July 2001.

Bahcheli, Tozun, *Greek-Turkish Relations Since 1955*, Boulder, CO: Westview Press, 1990.

_____, "Turkey's Cyprus Challenge," in Dimitris Keridis and Dimitrios Triantaphyllou, eds., *Greek-Turkish Relations in an Era of Globalization*, Dallas, VA: Brassey's, 2001, pp. 216–217.

Bal, Idris, *Turkey's Relationship with the West and the Turkic Republics*, Burlington, VT: Ashgate Publishing Company, 2000.

Bal, Idris, and Cengiz Basak Bal, "Rise and Fall of Elchibey and Turkey's Central Asian Policy," *Dis Politika*, No. 3-4, 1998, pp. 42–56.

Barchard, David, *Building a Partnership: Turkey and the European Union*, Istanbul: Turkish Economic and Social Studies Foundation, 2000.

Barkey, Henri J., "The Silent Victor: Turkey's Role in the Gulf War," in Efraim Karsh, ed., The Iran-Iraq War: Impact and Implications, London: Macmillan, 1989, pp. 133–153.

_____, "The Struggles of a Strong State," *The Journal of International Affairs*, Vol. 54, No. 1, Fall 2000, pp. 87–105.

_____, ed., *Reluctant Neighbor: Turkey's Role in the Middle East*, Washington, D.C.: U.S. Institute of Peace, 1996.

Barkey, Henri J., and Graham E. Fuller, *Turkey's Kurdish Question*, Lanham, MD: Rowman and Littlefield, 1998.

Bengio, Ofra, and Gencer Ozcan, *Arab Perceptions of Turkey and its Alignment with Israel*, Tel Aviv: BESA Center for Strategic Studies, 2001.

Binder, Leonard, ed., *Ethnic Conflict and International Politics in the Middle East*, Gainsville, FL: University Press of Florida, 1999.

Birand, Mehmet Ali, *Shirts of Steel: An Anatomy of the Turkish Armed Forces*, London: I. B. Tauris, 1991.

_____, "Turkey and the Davos Process," in Dimitri Constas, ed., *The Greek-Turkish Conflict in the 1990s*, New York: St. Martin's Press, 1991, pp. 27–39.

Blackwill, Robert D., and Michael Stürmer, eds. *Allies Divided: Transatlantic Policies for the Greater Middle East*, Cambridge, MA: The MIT Press, 1997.

Brewin, Christopher, *The European Union and Cyprus*, Huntingdon, England: The Eothen Press, 2000.

Brown, James, *Delicately Poised Allies: Greece and Turkey*, London: Brassey's, 1991.

Brown, L. Carol, ed., *Imperial Legacy. The Ottoman Imprint on the Balkans and the Middle East*, New York: Columbia University Press, 1996,

Bugra, Aysa, *State and Business in Modern Turkey: A Comparative Study*, Albany, NY: State University of New York Press, 1994.

Candar, Cengiz, "Some Turkish Perspectives on the United States and American Policy Toward Turkey," in Morton Abramowitz, ed., *Turkey's Transformation and American Policy*, New York: The Century Foundation Press, 2000, pp. 123–152.

_____, "Atatürk's Ambiguous Legacy," *The Wilson Quarterly*, Vol. XXIV, No. 4, Autumn 2000, pp. 88–96.

Celac, Sergiu, Michael, Emerson, and Nathalie Focci, *A Stability Pact for the Caucasus*, Brussels: Center for European Policy, May 2000.

Chace, Robert S., Emily Hill, and Paul Kennedy, "Pivotal States and U.S. Strategy," *Foreign Affairs*, Vol. 75, No. 1, January/February 1996, pp. 33–51.

Chace, Robert, et al., eds. *The Pivotal States*, New York: Norton, 1999.

Clogg, Richard, "Greek-Turkish Relations in the Post-1974 Period," in Dimitri Constas, ed., *The Greek-Turkish Conflict in the 1990s*, New York: St. Martin's Press, 1991, pp. 12–23.

Cohen, Ariel, "The 'New Great Game': Pipeline Politics in Eurasia," *European Studies*, Vol. 3, No. 1, Spring 1996, pp. 2–15.

Commission Opinion on Turkey's Request for Accession to the Community [SEC (89) 2290 fin./2], Brussels, December 20, 1989.

Constas, Dimitri, ed., *The Greek-Turkish Conflict in the 1990s*, New York: St. Martin's Press, 1991.

Cornell, Svante E., "Iran and the Caucasus," *Middle East Policy*, Vol. 5, No. 4, January 1998, pp. 51–67.

_____, "Turkey and the Conflict in Nagorno-Karabakh: A Delicate Balance," *Middle Eastern Studies*, Vol. 34, No. 1, January 1998, pp. 51–72.

_____, "Geopolitics and Strategic Alignments in the Caucasus and Central Asia," *Perceptions*, Vol. IV, No. 2, June–August 1999, pp. 100–125.

_____, "Turkey: Return to Stability?" *Middle Eastern Studies*, Vol. 35, No. 4, October 1999, pp. 228–231.

_____, "The Kurdish Question in Turkish Politics," *Orbis*, Vol. 45, No. 1, Winter 2000, pp. 31–46.

Couloumbis, Theodore A., *The United States, Greece and Turkey: The Troubled Triangle*, New York: Praeger, 1983.

Crawshaw, Nancy, *The Cyprus Revolt: An Account of the Struggle for Union with Greece*, London: George Allen & Unwin, 1978.

Demetriou, Madeleine, "On the Long Road to Europe and the Short Path to War: Issue-Linkage Politics and the Arms Build-Up in Cyprus," *Mediterranean Politics*, Vol. 3, No. 3, Winter 1998, pp. 38–51.

Dodd, Clement H., *The Cyprus Imbroglio*, Huntingdon, England: The Eothen Press, 1998.

_____, *Cyprus: The Need for New Perspectives*, Huntingdon, England: The Eothen Press, 1999.

_____, "A Historical Overview," in Clement H. Dodd, *Cyprus: The Need for New Perspectives*, Huntingdon, England: The Eothen Press, 1999.

_____, ed., *Turkish Foreign Policy. New Prospects*, Huntingdon, England: The Eothen Press, 1992.

Dunlop, John, "Russia Under Putin: Reintegrating the Post-Soviet Space," *Journal of Democracy*, Vol. 11, No. 3, 2000, pp. 39–47.

Eisenstadt, Michael, "Preparing for a Nuclear Breakout in the Middle East," *Policywatch*, No. 550, parts I and II, Washington, D.C.: The Washington Institute for Near East Policy, August 2001.

Ergil, Dogu, "Identity Crisis and Political Instability in Turkey," *The Journal of International Affairs*, Vol. 54, No. 1, Fall 2000, pp. 43–62.

Fedorov, Yuri E., "The Putin Factor in the Caspian," *Private View*, Autumn 2000, pp. 54–60.

Field, James A., *America and the Mediterranean World 1776–1882*, Princeton, NJ: Princeton University Press, 1969.

Finkel, Andrew, "Who Guards the Turkish Press? A Perspective on Press Corruption in Turkey," *The Journal of International Affairs*, Vol. 54, No. 1, Fall 2000, pp. 147–166.

Forsythe, Rosemary, "The Politics of Oil in the Caucasus and Central Asia," *Adelphi Paper 300*, London: International Institute for Strategic Studies, 1996.

Frey, Frederick, *The Turkish Political Elite*, Cambridge, MA: MIT Press, 1995.

Fuller, Graham E., "Turkey's New Eastern Orientation," in Graham E. Fuller and Ian O. Lesser, *Turkey's New Geopolitics: From the Balkans to Western China*, Boulder, CO: Westview/RAND, 1993.

Fuller, Graham E., and Ian O. Lesser, *Turkey's New Geopolitics: From the Balkans to Western China*, Boulder, CO: Westview/RAND, 1993.

Gellner, Ernst, *Encounters with Nationalism*, Oxford: Blackwell, 1994, pp. 81–91.

Gervasi, Frank, *Thunder over the Mediterranean*, New York: David McKay, 1975.

Gürbey, Gülistan, "Der Fall Öcalan und die türkisch-griechische Krise: Alte Drohungen oder neue Eskalation?" *Südosteuropa Mitteilungen*, Nr. 2, 1999, pp. 122–136.

Hale, William, *Turkish Foreign Policy 1774–2000*, London: Frank Cass, 2000.

_____, "Economic Issues in Turkish Foreign Policy," in Alan Makovsky and Sabri Sayari, eds., *Turkey's New World*, Washington,

D.C.: The Washington Institute for Near East Policy, 2000, pp. 20–38.

Harris, George H., *Troubled Alliance: Turkish-American Problems in Historical Perspective*, Washington, D.C.: American Enterprise Institute, 1972.

Heffner, Robert W., ed., *Democratic Civility*, New Brunswick, NJ: Transition, 1998.

Henze, Paul B., *Turkey and Armenia: Past Problems and Future Prospects*, Santa Monica, CA: RAND, 1996.

_____, "Turkey and the Caucasus," *Orbis*, Vol. 45, No. 1, Winter 2001, pp. 81–91.

Heper, Metin, "Islam and Democracy in Turkey: Toward a Reconciliation?" *The Middle East Journal*, Vol. 51, No. 1, Winter 1997, pp. 32–45.

Herzig, Edmund, *Iran and the Former Soviet South*, London: Royal Institute of International Affairs, 1995.

IISS Military Balance, 2000–2001, London: International Institute for Strategic Studies, 2000.

Inbar, Efraim, "The Strategic Glue in the Israeli-Turkish Alignment," in Barry Rubin and Kemal Kirişçi, eds., *Turkey in World Politics*, Boulder, CO: Lynne Rienner, 2001, pp. 115–126.

Jenkins, Gareth, "Turkey and EU Security: Camouflage or Criterion for Candidacy?" *Security Dialogue*, Vol. 32, No. 2, June 2001, pp. 269–272.

_____, *Context and Circumstance: The Turkish Military and Politics*, Adelphi Paper 337, London: International Institute for Strategic Studies, 2001.

Kahraman, Sevilay Elgün, "Rethinking Turkey–European Union Relations in Light of Enlargement," *Turkish Studies*, Vol. 1, No. 1, Spring 2000, pp. 1–20.

Kalicki, Jan, "Caspian Energy at the Crossroads," *Foreign Affairs*, Vol. 80, No. 5, September–October 2001, pp. 120–134.

Karpat, Kemal, ed., *The Turks of Bulgaria: The History, Culture, and Political Fate of a Minority*, Istanbul: The ISIS Press, 1990.

Karsh, Efraim, ed., *The Iran-Iraq War: Impact and Implications*, London: Macmillan, 1989.

Kasaba, Resat, "Cohabitation? Islamist and Secular Groups in Modern Turkey," in Robert W. Heffner ed., *Democratic Civility*, New Brunswick, NJ: Transition 1998.

Keridis, Dimitris, and Dimitrios Triantaphyllou, *Greek-Turkish Relations in an Era of Globalization*, Dallas, VA: Brassey's 2001.

Khalilzad, Zalmay, Ian O. Lesser, and F. Stephen Larrabee, *The Future of Turkish-Western Relations: Toward a Strategic Plan*, Santa Monica, CA: RAND, 2000.

Kinzer, Stephen, *Crescent & Star*, New York: Farrar, Strauss and Giroux, 2001.

Kirişçi, Kemal, "Turkey and the Muslim Middle East," in Alan Makovsky and Sabri Sayari, eds., *Turkey's New World*, Washington, D.C.: The Washington Institute for Near East Policy, 2000, pp. 39–58.

_____, "Disaggregating Turkish Citizenship and Immigration Practices," *Middle Eastern Studies*, Vol. 26, No. 3, July 2000, pp. 14–17.

_____, "The Future of Turkish Policy Toward the Middle East," in Barry Rubin and Kemal Kirişçi, eds., *Turkey in World Politics: An Emerging Multiregional Power*, London: Lynne Rienner Publishers, 2001, pp. 93–114.

Klare, Michael T., *Resource Wars*, New York: Henry Holt and Company, 2001.

Koliopoulos, John, and Thanos Veremis, *Greece, the Modern Sequel*, New York: New York University Press, 2002.

Kramer, Heinz, *Die Europäische Gemeinschaft und die Turkei: Entwicklung, Probleme und Perspektiven einer schwierigen Partnerschaft*, Baden-Baden: Nomos Verlagsgesellschaft, 1988.

_____, "Turkey's Relations with Greece: Motives and Interests," in Dimitri Constas, ed., *The Greek-Turkish Conflict in the 1990s*, New York: St. Martin's Press, 1991, pp. 57–72.

_____, "Turkey and the European Union: A Multi-Dimensional Relationship," in Vojtech Mastny and R. Craig Nation, eds., *Turkey Between East and West: New Challenges for a Rising Regional Power*, Boulder, CO: Westview Press, 1996, pp. 203–232.

_____, *A Changing Turkey: The Challenges to Europe and the U.S.*, Washington, D.C.: The Brookings Institution, 2000.

Kugler, Richard L., and Ellen L. Frost, eds., *The Global Century, Globalization and National Security, Volume II*, Washington, D.C.: National Defense University Press, 2001.

Kuniholm, Bruce R., *The Origins of the Cold War in the Near East: Great Power Conflict and Diplomacy in Iran, Turkey and Greece*, Princeton, NJ: Princeton University Press, 1980.

Larrabee, F. Stephen, "U.S. and European Policy Toward Turkey and the Caspian Basin," in Robert D. Blackwill and Michael Stürmer, eds., *Allies Divided: Transatlantic Policies for the Greater Middle East*, Cambridge, MA: The MIT Press, 1997, pp. 143–173.

_____, "Turkish Foreign and Security Policy: New Dimensions and New Challenges," in Zalmay Khalilzad, Ian O. Lesser, and F. Stephen Larrabee, *The Future of Turkish-Western Relations: Toward a Strategic Plan*, Santa Monica, CA: RAND, 2000, pp. 21–51.

_____, "Russia and Its Neighbors: Integration or Disintegration," in Richard L. Kugler and Ellen L. Frost, eds., *The Global Century, Globalization and National Security, Volume II*, Washington, D.C.: National Defense University Press, 2001, pp. 859–874.

Lesser, Ian O., *Bridge or Barrier? Turkey and the West After the Cold War*, Santa Monica, CA: RAND, 1992.

_____, "Beyond Bridge or Barrier: Turkey's Evolving Security Relations with the West," in Alan Makovsky and Sabri Sayari, eds., *Turkey's New World: Changing Dynamics in Turkish Foreign Policy*, Washington, D.C.: The Washington Institute for Near East Policy, 2000, pp. 203–221.

_____, "Turkey in a Changing Security Environment," *The Journal of International Affairs*, Vol. 54, No. 1, Fall 2000, pp. 183–198.

_____, *Strong States, Difficult Choices: Mediterranean Perspectives on Integration and Globalization*, Washington, D.C.: National Intelligence Council/RAND, 2001.

_____, *Turkey, Greece and the U.S. in a Changing Strategic Environment: Testimony Before the House International Relations Committee, Subcommittee on Europe*, Santa Monica, CA: RAND, June 2001.

Lesser, Ian O., and Ashley J. Tellis, *Southern Exposure: Proliferation Around the Mediterranean*, Santa Monica, CA: RAND, 1996.

Lesser, Ian O., Jerrold D. Green, F. Stephen Larrabee, and Michele Zanini, *The Future of NATO's Mediterranean Initiative: Prospects and Next Steps*, Santa Monica, CA: RAND, 2000.

Lesser, Ian O., F. Stephen Larrabee, Michele Zanini, and Katia Vlachos-Dengler, *Greece's New Geopolitics*, Santa Monica, CA: RAND, 2001.

Lewis, Bernard, *The Emergence of Modern Turkey*, London: Oxford University Press, 1961.

Lieven, Anatole, "The (Not So) Great Game," *The National Interest*, Winter 1999/2000, pp. 69–80.

Lustick, Ian, "Hegemony and the Riddle of Nationalism," in Leonard Binder, ed., *Ethnic Conflict and International Politics in the Middle East*, Gainesville, FL: University Press of Florida, 1999.

Makovsky, Alan O., "Marching in Step, Mostly!" *Private View*, Spring 1999, Vol. 3, No. 7, p. 38.

_____, "The New Activism in Turkish Foreign Policy," *SAIS Review*, Vol. 19, No. 1, Winter–Spring 1999, pp. 92–113.

_____, "Turkey's Nationalist Moment," *The Washington Quarterly*, Vol. 22, No. 4, Autumn 1999, pp. 159–166.

_____, "Turkish-Israeli Ties in the Context of Arab-Israeli Tension," *Policywatch*, No. 502, Washington, D.C.: The Washington Institute for Near East Policy, November 10, 2000.

_____, "Turkey and the Bush Administration: The Question Marks," *Policywatch*, No. 527, Washington. D.C.: The Washington Institute for Near East Policy, March 30, 2001.

Makovsky, Alan, and Sabri Sayari, eds., *Turkey's New World*, Washington, D.C.: The Washington Institute for Near East Policy, 2000.

Mango, Andrew, *Turkey: The Challenge of a New Role*, Westport, CT: Praeger, 1994.

_____, *Atatürk: The Biography of the Founder of Modern Turkey*, New York: Overlook Press, 1999.

_____, "Atatürk and the Kurds," *Middle Eastern Studies*, Vol. 35, No. 4, October 1999, pp. 1–25.

_____, "Reflections on the Atatürkist Origins of Turkish Foreign Policy and Domestic Linkages," in Alan Makovsky and Sabri Sayari, eds., *Turkey's New World: Changing Dynamics in Turkish Foreign Policy*, Washington, D.C.: The Washington Institute for Near East Policy, 2000, pp. 9–19.

Mastny, Vojtech, and R. Craig Nation, eds., *Turkey Between East and West: New Challenges for a Rising Regional Power*, Boulder, CO: Westview Press, 1996,

Mayall, Simon V., *Turkey: Thwarted Ambition*, McNair Paper No. 56, Washington, D.C.: National Defense University, 1997.

Micu, Nicolae, "Black Sea Economic Cooperation (BSEC) as a Confidence-Building Measure," *Perceptions*, Vol. 1, No. 4, December–February 1996/97, pp. 68–75.

Missiroli, Antonio, "EU-NATO Cooperation in Crisis Management: No Turkish Delight for ESDP," *Security Dialogue*, Vol. 33, No. 1, March 2002, pp. 9–26.

Mufti, Malik, "Daring and Caution in Turkish Foreign Policy," *The Middle East Journal*, Vol. 52, No. 1, Winter 1998, pp. 32–50.

Müftüler-Bac, Meltem, "The Never Ending Story: Turkey and the European Union," *Middle Eastern Studies*, Vol. 34, No. 4, October 1998, pp. 240–258.

_____, "Through the Looking Glass: Turkey in Europe," *Turkish Studies*, Vol. 1, No. 1, Spring 2000, pp. 21–35.

_____, "Turkey's Role in the EU's Security and Foreign Policies," *Security Dialogue*, Vol. 31, No. 4, December 2000, pp. 489–502.

Nachmani, Amikam, "Turkey and the Middle East," *BESA Security and Policy Studies*, No. 42, 1999, p. 3.

Neumann, Iver B., *Uses of the Other: The "East" in European Identity Formation*, Minneapolis, MN: University of Minnesota Press, 1999.

Nodia, Ghia, "Turmoil and Stability in the Caucasus: Internal Developments and External Influence," presentation at the conference "Prospects for Regional and Transregional Cooperation and the Resolution of Conflicts," Yerevan, Armenia, September 27–28, 2000, pp. 85–94.

Olcott, Martha Brill, "The Caspian's False Promise," *Foreign Policy*, No. 111, Summer 1998, pp. 95–113.

Olson, Robert, *The Emergence of Kurdish Nationalism and the Sheikh Sa'id Rebellion, 1880–1925*, Austin, TX: University of Texas Press, 1989.

Onis, Ziya, "Neo-Liberal Globalization and the Democracy Paradox: The Turkish General Elections of 1999," *The Journal of International Affairs*, Vol. 54, No. 1, Fall 2000, pp. 283–306.

_____, "Luxembourg, Helsinki and Beyond: Towards an Interpretation of Recent Turkey-EU Relations," *Government and Opposition*, Vol. 35, No. 4, Autumn 2000, pp. 403–483.

Öymen, Onur, *The Turkish Challenge*, Nicosia (Northern Cyprus): Rustem, 2000.

Özer, Ercan, "Concept and Prospects of the Black Sea Economic Cooperation," *Foreign Policy Review*, Vol. XX, No. 1-2, 1996, pp. 75–106.

Park, William, "Turkey's European Union Candidacy: From Luxembourg to Helsinki—to Ankara?" *Mediterranean Politics*, Vol. 5, No. 3, Autumn 2000, pp. 31–53.

Polyviou, Polyvios, *Cyprus: Conflict and Negotiations, 1960–1980*, London: Duckworth, 1980.

Pope, Nicole, and Hugh Pope, *Turkey Unveiled: Atatürk and After*, London: Murray, 1997.

Poulton, Hugh, *The Balkans: Minorities and States in Conflict*, London: Minority Rights Publications, 1991.

_____, *Top Hat, Grey Wolf and Crescent*, New York: NYU Press, 1997.

Psomiades, Harry J., *The Eastern Question: The Last Phase*, New York: Pella, 2000.

Radu, Michael, "The Rise and Fall of the PKK," *Orbis*, Vol. 45, No. 1, Winter 2001, pp. 47–63.

Randal, Jonathan C., *After Such Knowledge, What Forgiveness?"* Boulder, CO: Westview Press, 1999.

Rhein, Eberhard, "Europe and the Greater Middle East," in Robert D. Blackwill and Michael Stürmer, eds., *Allies Divided: Transatlantic Policies for the Greater Middle East*, Cambridge, MA: The MIT Press, 1997, pp. 41–59.

Robins, Philip, *Turkey and the Middle East*, New York: Council on Foreign Relations Press, 1991.

_____, "Between Sentiment and Self-Interest: Turkey's Policy Toward Azerbaijan and the Central Asian States," *The Middle East Journal*, Vol. 47, No. 4, Autumn 1993, pp. 593–610.

_____, "Turkey and the Kurds: Missing Another Opportunity?" in Morton Abramowitz, ed., *Turkey's Transformation and American Policy*, New York: The Century Foundation Press, 2000, pp. 61–93.

Rouleau, Eric, "Turkey's Dream of Democracy," *Foreign Affairs*, Vol. 79, No. 6, November/December 2000, pp. 100–114.

Rubin, Barry, *Istanbul Intrigues*, New York: McGraw-Hill, 1989.

Rubin, Barry, and Kemal Kirişçi, eds., *Turkey in World Politics*, Boulder, CO: Lynne Rienner, 2001.

Rumer, Eugene, *Dangerous Drift: Russia's Middle East Policy*, Washington, D.C.: The Washington Institute for Near East Policy, October 2000.

Salt, Jeremy, "Nationalism and the Rise of Muslim Sentiment," *Middle Eastern Studies*, Vol. 31, No. 1, January 1995, pp. 13–27.

Sayari, Sabri, "Turkish Foreign Policy in the Post–Cold War Era: The Challenges of Multi-Regionalism," *The Journal of International Affairs*, Vol. 54, No. 1, Fall 2000, pp. 169–182.

Sezer, Duygu Bazoglu, "Turkish-Russian Relations from Adversary to 'Virtual Rapprochement,'" in Alan Makovsky and Sabri Sayari, eds., *Turkey's New World: Changing Dynamics in Turkish Foreign Policy*, Washington, D.C.: The Washington Institute for Near East Policy Studies, 2000, pp. 92–115.

_____, "Turkish-Russian Relations: The Challenge of Reconciling Geopolitical Competition and Economic Partnership," *Turkish Studies*, Vol. 1, No. 1, Spring 2000, pp. 59–82.

_____, "Turkish-Russian Relations a Decade Later: From Adversary to Managed Competition," *Perceptions*, Vol. VI, No. 1, March–May 2001, pp. 79–98.

Shaffer, Brenda, *Partners in Need: The Strategic Relationship of Russia and Iran*, Washington, D.C.: The Washington Institute for Near East Policy, May 2001.

Smith, Anthony D., *National Identity*, Reno, NV: University of Nevada Press, 1991.

Sokolsky, Richard, and Tanya Charlick-Paley, *NATO and Caspian Security: A Mission Too Far?* Santa Monica, CA: RAND, 1999.

A Stability Pact for the Caucasus, Brussels: Center for European Policy, May 2000.

Starr, S. Frederick, "Power Failure in the Caspian," *The National Interest*, No. 47, Spring 1997, pp. 20–31.

Stavrinides, Zenon, "Greek Cypriot Perceptions," in Clement H. Dodd, ed., *Cyprus: The Need for New Perspectives*, Huntingdon, England: The Eothen Press, 1999, pp. 54–96.

Stearns, Monteagle, *Entangled Allies: U.S. Policy Toward Greece, Turkey and Cyprus*, New York: The Council on Foreign Relations, 1992.

Tocci, Natalie, *21st Century Kemalism: Redefining Turkey-EU Relations in the Post Helsinki Era*, Brussels: Center for European Policy (no date).

Todorova, Maria, "The Ottoman Legacy in the Balkans," in L. Carol Brown, ed., *Imperial Legacy. The Ottoman Imprint on the Balkans and the Middle East*, New York: Columbia University Press, 1996, pp. 45–77.

Toward Calmer Waters: A Report on Relations Between Turkey and the European Union, The Hague: Advisory Council on International Affairs, July 1999.

Triantaphyllou, Dimitrios, "Further Turmoil Ahead?" in Dimitris Keridis and Dimitrios Triantaphyllou, *Greek-Turkish Relations in an Era of Globalization*, Dallas, VA: Brassey's, 2001, pp. 73–74.

Türkes, Mustafa, "The Balkan Pact and Its Immediate Implications for the Balkan States, 1930–1934," *Middle Eastern Studies*, Vol. 30, No. 1, January 1994, pp. 123–144.

"Turkey and the Kurds: Into a New Era," *IISS Strategic Comments*, Vol. 5, Issue 3, April 1999.

"Turkey's Divided Islamists," *IISS Strategic Comments*, Vol. 6, Issue 3, April 2000.

Turkey's Window of Opportunity: The Demographic Transition Process and Its Consequences, Istanbul: TUSIAD, 1999.

Turkish Economy 1999–2000, Istanbul: TUSIAD, July 2000.

Ulusoy, Hason, "A New Formation in the Black Sea: BLACK-SEAFOR," *Perceptions*, Vol. VI, No. 4, December 2001–February 2002, pp. 97–106.

U.S. Department of State, *Turkey: Country Report on Human Rights Practices for 1998*, Washington, D.C., 1999.

Vali, Ferenc A., *Bridge Across the Bosporus: The Foreign Policy of Turkey*, Baltimore, MD: The Johns Hopkins University Press, 1971.

Valinakis, Yannis, *Greece's Security in the Post–Cold War Era*, Ebenhausen: Stiftung Wissenschaft und Politik, S-394, April 1994.

_____, "The Black Sea Region: Challenges and Opportunities for Europe," *Chaillot Papers 36*, Paris: West European Union Institute for Security Studies, July 1999.

Williams, Paul, "Turkey's H2O Diplomacy in the Middle East," *Security Dialogue*, Vol. 32, No. 1, March 2001, pp. 27–40.

Wilson, Andrew, *The Aegean Dispute*, London: International Institute for Strategic Studies, 1980.

Winrow, Gareth, *Turkey in Post-Soviet Central Asia*, London: Royal Institute of International Affairs, 1995.

_____, "Turkey and the Newly Independent States of Central Asia and the Caucasus," in Barry Rubin and Kemal Kirişçi, eds., *Turkey in World Politics: An Emerging Multiregional Power*, London: Lynne Rienner Publishers, 2001, pp. 183–184.

Yavuz, H. Hakan, "Towards an Islamic Liberalism: The Nurcu Movement and Fethullah Gülen," *The Middle East Journal*, Vol. 53, No. 4, Autumn 1999, pp. 584–605.

_____, "Cleansing Islam from the Public Sphere," *The Journal of International Affairs*, Vol. 54, No. 1, Fall 2000, pp. 21–42.

F. Stephen Larrabee is a senior staff member at RAND in Washington, D.C., and holds the RAND Corporate Chair in European Security. He has a Ph.D. in political science from Columbia University and has taught at Columbia University, Cornell University, New York University, the Paul Nitze School of Advanced International Studies (SAIS), Georgetown University, and the University of Southern California. Before joining RAND, he served as Vice President and Director of Studies of the Institute of East-West Security Studies in New York from 1983 to 1989 and was a distinguished Scholar in Residence at the institute from 1989 to 1990. From 1978–1981, Dr. Larrabee served on the U.S. National Security Council staff in the White House as a specialist on Soviet–East European affairs and East-West political-military relations. He is coeditor (with Zalmay Khalilzad and Ian O. Lesser) of *The Future of Turkish-Western Relations*; coeditor (with David Gompert) of *America and Europe: A Partnership for a New Era* (1997); author of *East European Security After the Cold War* (1994); editor of *The Volatile Powder Keg: Balkan Security After the Cold War* (1994); coeditor (with Robert Blackwill) of *Conventional Arms Control and East-West Security* (1989); and editor of *The Two German States and European Security* (1989).

Ian O. Lesser is Vice President, Director of Studies, at the Pacific Council on International Policy in Los Angeles. He came to the Council from RAND, where he was a senior political scientist specializing in strategic studies and international policy. In 1994–1995, he was a member of the State Department's Policy Planning Staff, where his responsibilities included southern Europe, Turkey, North Africa, and the Middle East peace process. Earlier in his career, he was deputy director of the Political-Military Studies Program at the Center for Strategic and International Studies, senior fellow at the Atlantic Council, and staff consultant at International Energy Associates. His recent publications include *NATO Looks South* (2000); *Countering the New Terrorism* (1999); and *Sources of Conflict in the 21st Century* (1998); and he is a frequent commentator for television, radio, and print media. Dr. Lesser is a graduate of the University of Pennsylvania, the London School of Economics, and The Fletcher School of Law and Diplomacy. He received his D. Phil in international politics from Oxford University. He is a member of the Council on Foreign Relations and the International Institute for Strategic Studies.